新楼盘 42
NEWHOUSE 图解地产与设计

总部科技园

中国林业出版社

AECF
AUROS ESPACES CONCEPTION (FRANCE)

法国颐朗联合建筑设计有限公司

Phone: 021-65909515
E-mail: www.YL-AECF.com
Websie: YL_AECF@163.com

| 住宅地產 | 商業地產 | 旅游地產 | 度假酒店 | 市政規劃 | 校園規劃 | 公園規劃 | 區域規劃 |

法國建築師協會會員單位

美國靈頓建築景觀設計有限公司是專業從事城市規劃和城市設計、風景區與公園景觀規劃、風景旅游渡假景觀、酒店環境景觀、高尚住宅小區景觀及工程設計的國際知名公司。深圳靈頓建築景觀設計有限公司是其專爲中國地區設置的設計顧問公司，公司以國外多位設計師爲公司主幹，美國公司作爲主要規劃整體和宏觀控制，以共同協作，共同致力于同一項目上，使設計工程在整體規劃及局部處理上都得到精心設計。公司以全新的理念引導市場，以專業的服務介入市場，以全程的服務方式開拓市場，使公司得到穩步和迅速的發展，業務範圍不斷擴大。尤其在高尚住宅小區景觀設計中取的優异的成績，其設計工程多次獲國家和地區的獎勵。

http://www.szld2005.com　　　中國 深圳 福田區 紅荔路花卉世界313號

TEL: (0755) 8621 0770　　FAX: (0755) 8621 0772　　P.C: 518000　　EMAIL: szld2000@163.com　　179049195@QQ.com

美國靈頓建築景觀設計有限公司　　深圳靈頓建築景觀設計有限公司　　雲南靈頓園林綠化工程有限公司

- 综合策划、规划、建筑、景观于一体化的关联性设计
- 强调因时、因地、因人制宜的多目标体系的系统性设计
- 尊崇理性而随机、行云流水、动态平衡的非线性设计

广州市金冕建筑设计有限公司

(King Made Group) 始创于2002年10月

公司追求以——

实是 **G**enuine
理性 **R**ational
原创 **O**riginal
时进 **U**pdating
唯美 **P**erfect

虚位以待：设计总监、主任建筑师、建筑师、规划师、景观设计师、工程师

广州·南宁·长沙

广州总公司：珠江新城 IFC(西塔)21层 06-08室 Tel：020-88832190
南宁分公司：民族大道 127号铂宫·国际 12层 Tel：0771-5736535
长沙分公司：岳麓区岳麓大道311号金麓商务广场 5栋 5楼 Tel：0731-89875038

▼ 三江华远浔江龙湾小区
▼ 佛山桂殿兰宫小区

▲ 山东烹饪学院
▲ 保定燕赵国际城市综合体
▲ 毕节大众国际花园

▲ 南宁太阳岛小区
▲ 增城荔城花园
◀ 王志纲丽江书院

▲ 常德西城中环城市综合体
◀ 梅州山水城项目
▶ 柳州山水一号项目

Web: http://www.kingmade.com
Tel: 020-88832190 88832191
Add: 广州市天河区珠江西路5号
广州国际金融中心主塔写字楼第21层06-08单元

LUCAS 奥德景观
DESIGN GROUP

西安阳光城-西西安小镇

深圳市奥德景观规划设计有限公司

简介

公司坐落于著名的蛇口湾畔，深圳最有影响力的创意设计基地：南海意库；
公司前身为深圳市卢卡斯景观设计有限公司，是由2003年成立于香港的卢卡斯联盟（香港）国际设计有限公司在世界设计之都：深圳设立的中国境内唯一公司。
于2012年1月获得中华人民共和国国家旅游局正式认定：旅游规划设计乙级设计资质。

公司专注于

居住区景观与规划设计（含旅游地产）
商业综合体景观与规划设计（含购物公园、写字楼及创意园区）
城市规划及空间设计
市民公园设计
酒店与渡假村景观规划与设计
旅游策划及规划设计

公司目标

倚重当下的中国的文化渊源结合世界潮流，尊重地域情感，在中国打造具强烈地域特征的、风格化的、国际化的，具前瞻性、可再生的的城市景观、人居环境、风情渡假区及自然保护区。

地　址：深圳市南山区蛇口兴华路南海意库2栋410室
电　话：0755-86270761
传　真：0755-86270762
邮　箱：lucasgroup_lucas@163.com
网　址：www.lucas-designgroup.com

HPA
海波建筑设计事务所
Architecture.Planning.Engineering.Interiors

诚聘精英及应届毕业生

地址：上海中山西路 1279 弄 6 号 11 楼

邮编：200051

电话：(021) 51168290

邮箱：msheng@hpa.cn

网址：http:// www.hpa.cn

GVL 怡境景观
GREENVIEW LANDSCAPE DESIGN LIMITED

- 景观设计 — **L**andscape Design
- 旅游度假项目规划 — **R**esorts and Leisure Planning
- 市政项目规划 — **U**rban Planning Design
- 居住环境项目规划 — **C**ommunity Planning
- 公园及娱乐项目规划 — **P**arks and Entertainment Planning and Design

英国国家园景
工业协会海外会员
British Association of Landscape Industries, Overseas Full Member of BALI

美国景观设计师协会企业会员
Corporate Member Of American Society Of Landscape Architects

广州总部
地 址：珠江新城华夏路49号津滨腾越大厦南塔8楼
邮 编：510623
电 话：(020) 87690558　87695498
　　　　38032762　38032729
传 真：(020) 87697706
邮 箱：greenv@163.com
网 址：www.greenview.com.cn

香港
地 址：北角渣华道18号嘉汇商业大厦2106室
邮箱编号：070432
传 真：(852) 22934388

北京
地 址：朝阳区亚运村阳光广场B2-1701室
邮 编：100101
传 真：(010) 64977992
电 话：(010) 64975897

www.greenview.com.cn

@GVL怡境景观
http://weibo.com/gvlcn

前言 EDITOR'S NOTE

浅说总部科技园
Elementary Introduction of Sci—tech Park Architectures

随着全球经济竞争加剧，总部经济作为一种新的经济形态产生了，即某区域由于特有的资源优势吸引企业将总部在该区域集群布局。应用总部科技园建筑的大多是国家级大中型企业以及地区性的主导科技企业。

在提倡"人性化"、"生态化"生活及办公方式的今天，传统企业园区的盖办公楼、研发楼及生产厂房的模式已无法适应企业长期的发展，如何创造一个"便于人们工作及生活、休闲的有活力的空间"？这是设计师们必须思考的问题，也是总部科技园建筑的目的。

总部科技园这类建筑集齐了几大要素，首先，其运用了混合开发的概念，有些企业园区除了办公空间外，已具备商业、住宅、餐饮、健身娱乐及教育、金融等其他服务配套设施，其次，其融合了中国传统建筑的围合式空间与西方开放式街区的规划布局，既有临街建筑又有花园式办公建筑；再者，其注重统一形象的设计，包括标识、建筑立面和外部环境，造就了强大的企业向心力和令人羡慕的办公环境氛围。

设计总部科技园也许已经不仅仅是如何"建办公楼"的问题，还有涉及园区的产业定位、发展方向、经营模式等诸多方面的思量，真让设计师们或费神或兴奋。

With the intensifying competition of world economy, headquarter economy comes out as a new economic phenomenon, which refers to a certain area with resources advantage which attracts enterprises to settle their headquarters in groups. Enterprises which apply the architectural form of Sci—tech Parks are usually state—level large and medium—sized enterprises and leading sci—tech companies in a region.

Today, in the context of promotion for humane and ecological lifestyle and working style, the traditional development zones featuring constructing offices, research centers and workshops have long been legged behind and couldn't meet the long—term development needs of contemporary enterprises. How to create a dynamic space which provides convenience to people's working, living and leisure at the same time? This is a question worthy of thinking by all architects and also a purpose for construction of Sci—tech Parks.

There gather a few key factors for Sci—tech Park buildings. Firstly, under the concept of mixed use development, besides office area in some enterprise parks, there are even other service facilities, including residential, commercial, catering, entertainment and fitness, educational and financial facilities. Secondly, Sci—tech park is a mix of the enclosed space of Chinese traditional architectures and Western open block planning while street front buildings combine with garden—type office buildings. Thirdly, the unified designs, like logo, outside facade and surrounding environment all help forge a powerful centrifugal force and favorable office atmosphere.

Sci—tech park involves not only the problem how to build office buildings, but also the industrial orientation, developing direction and operational mode, etc, which really trouble but excite architects.

jiatu@foxmail.com

2012年　总第42期

面向全国上万家地产商决策层、设计院、建筑商、材料商、专业服务商的精准发行

指导单位 INSTRUCTION UNIT
亚太地产研究中心

出品人 PUBLISHER
杨小燕 YANG XIAOYAN

主编 CHIEF EDITOR
王志 WANG ZHI

副主编 ASSOCIATE EDITOR
熊冕 XIONG MIAN

编辑记者 EDITOR REPOTERS
唐秋琳 TANG QIULIN
钟梅英 ZHONG MEIYING
胡明俊 HU MINGJUN
康小平 KANG XIAOPING
胡荔丽 HU LILI
吴辉 WU HUI
曹丹莉 CAO DANLI
朱秋敏 ZHU QIUMIN
王盼青 WANG PANQING

设计总监 ART DIRECTORS
杨先周 YANG XIANZHOU
何其梅 HE QIMEI

美术编辑 ART EDITOR
詹婷婷 ZHAN TINGTING

国内推广 DOMESTIC PROMOTION
广州佳图文化传播有限公司

市场总监 MARKET MANAGER
周中一 ZHOU ZHONGYI

市场部 MARKETING DEPARTMENT
方立平 FANG LIPING
熊光 XIONG GUANG
王迎 WANG YING
杨先凤 YANG XIANFENG
熊灿 XIONG CAN
刘佳 LIU JIA

图书在版编目（CIP）数据

新楼盘.总部科技园：汉英对照/佳图文化主编.
—— 北京：中国林业出版社，2012.10
ISBN 978-7-5038-6801-6

Ⅰ.①新... Ⅱ.①佳... Ⅲ.①建筑设计 - 中国 - 现代 - 图集 Ⅳ.①TU206

中国版本图书馆CIP数据核字(2012)第190365号
出版：中国林业出版社
主编：佳图文化
责任编辑：李顺 许琳
印刷：利丰雅高印刷(深圳)有限公司

特邀顾问专家 SPECIAL EXPERTS (排名不分先后)

赵红红 ZHAO HONGHONG	陈航 CHEN HANG
王向荣 WANG XIANGRONG	范勇 FAN YONG
陈世民 CHEN SHIMIN	赵士超 ZHAO SHICHAO
陈跃中 CHEN YUEZHONG	孙虎 SUN HU
邓明 DENG MING	梅卫平 MEI WEIPING
冼剑雄 XIAN JIANXIONG	林世彤 LIN SHITONG
陈宏良 CHEN HONGLIANG	熊冕 XIONG MIAN
胡海波 HU HAIBO	周原 ZHOU YUAN
程大鹏 CHENG DAPENG	李焯忠 LI ZHUOZHONG
范强 FAN QIANG	原帅让 YUAN SHUAIRANG
白祖华 BAI ZUHUA	王颖 WANG YING
杨承刚 YANG CHENGGANG	周敏 ZHOU MIN
黄宇奘 HUANG YUZANG	王志强 WANG ZHIQIANG / DAVID BEDJAI
梅坚 MEI JIAN	陈英梅 CHEN YINGMEI
陈亮 CHEN LIANG	吴应忠 WU YINGZHONG
张朴 ZHANG PU	曾繁柏 ZENG FANBO
盛宇宏 SHENG YUHONG	朱黎青 ZHU LIQING
范文峰 FAN WENFENG	曹一勇 CAO YIYONG
彭涛 PENG TAO	冀峰 JI FENG
徐农思 XU NONGSI	滕赛岚 TENG SAILAN
田兵 TIAN BING	王毅 WANG YI
曾卫东 ZENG WEIDONG	陆强 LU QIANG
马素明 MA SUMING	徐峰 XU FENG
仇益国 QIU YIGUO	张奕和 EDWARD Y. ZHANG
李宝章 LI BAOZHANG	郑竞晖 ZHENG JINGHUI
李方悦 LI FANGYUE	刘海东 LIU HAIDONG
林毅 LIN YI	凌敏 LING MIN

编辑部地址：广州市海珠区新港西路3号银华大厦4楼
电话：020—89090386/42/49、28905912
传真：020—89091650

北京办：王府井大街277号好友写字楼2416
电话：010—65266908　　**传真**：010—65266908

深圳办：深圳市福田区彩田路彩福大厦B座23F
电话：0755—83592526　　**传真**：0755—83592536

协办单位 CO—ORGANIZER

 广州市金冕建筑设计有限公司　熊冕 总设计师
地址：广州市天河区珠江西路5号国际金融中心主塔21楼06—08单元
TEL：020—88832190　88832191
http://www.kingmade.com

AECF 上海颐朗建筑设计咨询有限公司　巴学天 上海区总经理
地址：上海市杨浦区大连路970号1308室
TEL：021—65909515　FAX：021—65909526
http://www.yl-aecf.com

WEBSITE COOPERATION MEDIA
网站合作媒体

SouFun 搜房网

副理事长单位 DEPUTY CHAIRMAN

华森建筑与工程设计顾问有限公司　邓明 广州公司总经理
地址：深圳市南山区滨海之窗办公楼6层
　　　广州市越秀区德政北路538号达信大厦26楼
TEL：0755—86126888　020—83276688
http://www.huasen.com.cn　E—mail:hsgzaa@21cn.net

广州瀚华建筑设计有限公司　冼剑雄 董事长
地址：广州市天河区黄埔大道中311号羊城创意产业园2—21栋
TEL：020—38031268　FAX：020—38031269
http://www.hanhua.cn
E—mail：hanhua—design@21cn.net

上海中建筑设计院有限公司　徐峰 董事长
地址：上海市浦东新区东方路989号中达广场12楼
TEL：021—68758810　FAX：021—68758813
http://www.shzjy.com
E—mail：csaa@shzjy.com

常务理事单位 EXECUTIVE DIRECTOR OF UNIT

深圳市华域普风设计有限公司　梅坚 执行董事
地址：深圳市南山区海德三道海岸城东座1306—1310
TEL：0755—86290985　FAX：0755—86290409
http://www.pofart.com

华通设计顾问工程有限公司
地址：北京市西城区西直门南小街135号西派国际C—Park3号楼
TEL：8610—83957395　8610—83957390
http://www.wdce.com.cn

天萌（中国）建筑设计机构　陈宏良 总建筑师
地址：广州市天河区员村四横路128号红专厂F9栋天萌建筑馆
TEL：020—37857429　FAX：020—37857590
http://www.teamer—arch.com

GVL国际怡境景观设计有限公司　彭涛 中国区董事及总监
地址：广州市珠江新城华夏路49号津滨腾越大厦南塔8楼
TEL：020—87690558　FAX：020—87697706
http://www.greenview.com.cn

天友建筑设计股份有限公司　马素明 总建筑师
地址：北京市海淀区西四环北路158号慧科大厦7F（方案中心）
TEL：010—88592005　FAX：010—88229435
http://www.tenio.com

R—LAND 北京源树景观规划设计事务所　白祖华 所长
地址：北京朝阳区朝外大街怡景园5—9B
TEL：010—85626992/3　FAX：010—85625520
http://www.ys—chn.com

奥雅设计集团　李宝章 首席设计师
深圳总部地址：深圳蛇口南海意象5栋302
TEL：0755—26826690　FAX：0755—26826694
http://www.aoya—hk.com

北京寰亚国际建筑设计有限公司　赵士超 董事长
地址：北京市朝阳区琨莎中心1号楼1701
TEL：010—65797775　FAX：010—84682075
http://www.hygjjz.com

广州山水比德景观设计有限公司　孙虎 董事总经理兼首席设计师
地址：广州市天河区珠江新城临江大道685号红专厂F19
TEL：020—37039822/823/825　FAX：020—37039770
http://www.gz—spi.com

奥森国际景观规划设计有限公司　李焯忠 董事长
地址：深圳市南山区南海大道粤海路动漫园7栋5楼
TEL：0755—26828246　86275795　FAX：0755—26822543
http://www.oc—la.com

广州市四季园林设计工程有限公司　原帅让 总经理兼设计总监
地址：广州市天河区龙怡路117号银汇大厦2505
TEL：020—38273170　FAX：020—86682658
http://www.gz—siji.com

深圳市雅蓝图景观设计工程有限公司　周敏 设计董事
地址：深圳市南山区南海大道2009号新能源大厦A座6D
TEL：0755—26650631/26650632　FAX：0755—26650623
http://www.yalantu.com

深圳市佰邦建筑设计顾问有限公司　迟春儒 总经理
地址：深圳市南山区兴工路8号美年广场1栋804
TEL：0755—86229594　FAX：0755—86229559
http://www.pba—arch.com

北京新纪元建筑工程有限公司　曾繁柏 董事长
地址：北京市海淀区小马厂6号华天大厦20层
TEL：010—63483388　FAX：010—63265003
http://www.bjxinjiyuan.com

北京博地澜屋建筑规划设计有限公司　曹一勇 总设计师
地址：北京市海淀区中关村南大街31号神舟大厦8层
TEL：010—68118690　FAX：010—68118691
http://www.buildinglife.com.cn

HPA上海海波建筑设计事务所　陈立波、吴海青 公司合伙人
地址：上海中山西路1279弄6号楼国峰科技大厦11层
TEL：021—51168290　FAX：021—51168240
http://www.hpa.cn

香港华艺设计顾问（深圳）有限公司　林毅 总建筑师
地址：深圳市福田区华富路航都大厦14、15楼
TEL：0755—83790262　FAX：0755—83790289
http://www.huayidesign.com

哲思（广州）建筑设计咨询有限公司　郑竞晖 总经理
地址：广州市天河区天河北路626号保利中宇广场A栋1001
TEL：020—38823593　FAX：020—38823598
http://www.zenx.com.au

理事单位 COUNCIL MEMBERS （排名不分先后）

广州柏澳景观设计有限公司　徐农思 总经理
地址：广州市天河区天园东路2191号时代新世界中心南塔2704室
TEL：020—87569202
http://www.bacdesign.com.cn

中房集团建筑设计有限公司　范强 总经理/总建筑师
地址：北京市海淀区百万庄建设部院内
TEL：010—68347818

北京奥思得建筑设计有限公司　杨承冈 董事总经理
地址：北京朝阳区东三环中路39号建外SOHO16号楼2903~2905
TEL：86—10—58692509/19/39　FAX：86—10—58692523

陈世民建筑师事务所有限公司　陈世民 董事长
地址：深圳市福田中心区益田路4068号卓越时代广场4楼
TEL：0755—88262516/429

广州嘉柯园林景观设计有限公司　陈航 执行董事
地址：广州市珠江新城华夏路49号津滨腾越大厦北塔506—507座
TEL：020—38032521/23　FAX：020—38032679
http://www.jacc—hk.com

侨恩国际（美国）建筑设计咨询有限公司
地址：重庆市渝北区龙湖MOCO4栋20—5
TEL：023—88197325　FAX：023—88197323
http://www.jnc—china.com

CDG国际设计机构　林世彤 董事长
地址：北京市长春路11号万柳亿城中心A座10/13层
TEL：010—58815603　58815633　FAX：010—58815637
http://www.cdgcanada.com

广州市圆美环境艺术设计有限公司　陈英梅 设计总监
地址：广州市海珠区宝岗大道杏坛大街56号二层之五
TEL：020—34267226　23353942　FAX：020—34267223
http://www.gzyuanmei.com

上海唯美景观设计工程有限公司　朱黎青 董事、总经理
地址：上海徐虹中路20号2—202室
TEL：021—61122209　FAX：021—61139033
http://www.wemechina.com

上海金创源建筑设计事务所有限公司　王毅 总建筑师
地址：上海市杨浦区黄兴路1858号701—703室
TEL：021—55062106　FAX：021—55062106—807
http://www.odci.com.cn

深圳灵顿建筑景观设计有限公司　刘海东 董事长
地址：深圳福田区红荔路花卉世界313号
TEL：0755—86210770　FAX：0755—86210772
http://www.szld2005.com

深圳市奥德景观规划设计有限公司　凌敏 董事总经理、首席设计师
地址：深圳市南山区蛇口海上世界南海意象2栋410#
TEL：0755—86270761　FAX：0755—86270762
http://www.lucas—designgroup.com

目录 CONTENTS

009	前言 EDITOR'S NOTE
014	资讯 INFORMATION

名家名盘 MASTER AND MASTERPIECE

- **018** 北京金科·王府：尊贵 典雅 自然 人文的法式宫廷社区
  FRENCH STYLE PALACE COMMUNITY: DIGNIFIED, ELEGANT, NATURAL AND HUMANISTIC

- **026** 西安紫薇·尚层：丰富而震撼的视觉 完整而统一的布局
  COLORFUL AND BREATH—TAKING VISION, INTEGRAL AND PERFECT LAYOUT

- **032** 上海新江湾城中建大公馆：英伦风格立面营造纯净人文社区
  BRITISH STYLE FACADE CREATES PURE CULTURAL COMMUNITY

专访 INTERVIEW

- **036** 设计要注重实现经济、功能与生态的和谐统一
  ——访北京寰亚国际建筑设计有限公司董事长 赵士超
  DESIGN SHOULD FOCUS ON THE HARMONY AND UNIFICATION OF ECONOMY, FUNCTION AND ECOLOGY

新景观 NEW LANDSCAPE

- **040** 南树峰药王养生谷规划设计：客家文化与药物养生的相融共生
  THE INTERGROWTH OF HAKKA CULTURE AND MEDICINE PRESERVATION

- **044** 中山万科城市风景：充满西班牙异域情调的浪漫文化之旅
  A ROMANTIC AND EXOTIC SPANISH CULTURAL TOUR

专题 FEATURE

- **052** 建筑：人与人之间的情感交流平台
  ——访北京博地澜屋建筑规划设计有限公司总设计师 曹一勇

BUILDING IS A PLATFORM FOR EMOTIONAL COMMUNICATION

056 ZOWONEN总部大楼：贯穿可持续理念的高效灵活空间
EFFICIENT AND FLEXIBLE SPACE WITH THE CONCEPT OF SUSTAINABILITY

062 科技园建设要把握好规划的前瞻性与设计的灵活性
——访ZENX INTERNATIONAL PTY LTD（哲思国际）张奕和博士
THE FORESIGHT OF PLANNING AND FLEXIBILITY OF DESIGN SHOULD BE HELD FOR SCI—TECH PARK PROJECTS

066 番禺节能科技园：生态、创意与新科技相融合的生态型科技园
INTEGRATION OF SUSTAINABILITY, CREATIVE DESIGN & INNOVATIVE TECHNOLOGY

072 CMT总部大楼：板式表皮覆盖下自由催生的标志性建筑
LANDMARK BUILDING INSPIRED BY FREEDOM UNDER THE SLAT COVER

080 杭州湾科技创业中心：简约而充满动感的后工业化园区
CONCISE AND DYNAMIC POSTINDUSTRIAL PARK

072

088 北京东亿国际传媒产业园：建筑铸就人与人之间情感交流的平台
BUILDINGS CREATE EMOTIONAL COMMUNICATION PLATFORMS AMONG PEOPLE

新特色 NEW CHARACTERISTICS

094 东方红郡：优雅 高贵 浪漫的法式风尚社区
ELEGANT, NOBLE, ROMANTIC FRENCH CLASSIC COMMUNITY

100 华彩国际公寓：汇聚都市核心区特征的高品质住区
CONVERGING CHARACTERISTICS OF URBAN CORE DISTRICTS

106 南宁太阳岛居住区规划设计方案：景观价值最大化运动
THE MAXIMUM OF LANDSCAPE VALUE

新空间 NEW SPACE

110 万科缤纷夏日样板房：激情 活力 简约 清新 明亮
PASSIONATE, ENERGETIC, SIMPLE, ELEGANT, BRIGHT

100

新创意 NEW IDEA

115 广州科学城科技人员集合住宅：视野开阔、充满活力的创意性居住空间
CREATIVE LIVING SPACE WITH WIDE VISION, FULL OF VIGOR

122 CAN BISA住宅：单一倾斜屋顶设计统一住宅空间
UNITY OF DWELLING SPACE WITH SINGLE—SLOPING ROOF DESIGN

114

商业地产
COMMERCIAL BUILDINGS

130 TARSU购物娱乐中心：交错重叠的多功能立体商业空间
STAGGERED MULTI—FUNCTIONAL COMMERCIAL SPACE

138 山西天美新天地：三位一体的高端城市综合体
TRINITARIAN HIGH—END HOPSCA

144 星月坊：定位于未来的餐饮功能商业项目
MULTIFUNCTIONAL COMMERCIAL PROJECT ORIENTED TO FUTURE

144

INFORMATION | 资讯/地产

统计局：8月CPI同比涨2.0% 居住价格升2.2%

9月9日，国家统计局发布的2012年8月居民消费价格主要数据显示，2012年8月份，全国居民消费价格总水平同比上涨2.0%。在各类商品及服务价格同比变动情况中，居住价格同比上涨2.2%。其中，住房租金价格上涨3%，水、电、燃料价格上涨3%，建房及装修材料价格上涨0.4%。

STATISTICAL BUREAU: CPI IN AUGUST GOES UP BY 2.0% YEAR—ON—YEAR, RESIDENCE PRICE INCREASES BY 2.2%

On September 9th, Consumer Price Index (CPI) of August released by National Statistical Bureau shows that national CPI climbs on a year—on—year ratio of 2.0% basis in August. Among all variation cases of various goods and service price, residence price has raised up by 2.2% year—on—year. Here, house rent goes up by 3%; water, electricity and fuel prices go up by 3%; prices of housing construction and fitment materials go up by 0.4%.

农产品批发市场和农贸市场免征房产税

财政部、国家税务总局近日联合下发通知，自2013年1月1日~2015年12月31日，对专门经营农产品的农产品批发市场、农贸市场使用的房产、土地，暂免征收房产税和城镇土地使用税；对同时经营其他产品的农产品批发市场和农贸市场使用的房产、土地，按其他产品与农产品交易场地面积的比例确定征免房产税和城镇土地使用税。

AGRICULTURAL PRODUCTS WHOLESALE MARKET AND FARM PRODUCT FAIR ARE FREE FOR BUILDING TAXES

Recently, Treasury Department and State Administration of Taxation both issue a notice that Agricultural Products Wholesale Market and Farm Product Fair specially run for farm products are free for building taxes and Tax on using urban land from January 1st, 2013 to December 31st, 2015. At the same time, t Agricultural Products Wholesale Market and Farm Product Fair specially run for other products will pay taxes according to the proportion of using land between farm products and other products.

住建部：前8月保障安居工程新开工650万套

住房和城乡建设部10日发布的数据显示，1~8月，全国城镇保障性安居工程新开工650万套，开工率为87%，基本建成420万套，完成投资8 200亿元。按照计划，今年我国将开工建设保障性安居工程700万套，比去年1 000万套的计划有所减少，但算上去年结转至今年的部分，所需的投资量仍较大。今年保障性安居工程的竣工任务为500万套。

BUILDING DEPARTMENT: HOUSING GUARANTEEING PROJECT STARTS 6.5 MILLION HOUSES IN THE FIRST EIGHT MONTHS

The Ministry of Housing and Urban—Rural Development declare that Housing guaranteeing projects start to work on 6.5 million houses with a ratio of 87% from January to August. 4.2 million houses have been finished in the investment of 820 billion yuan. As planned, China is going to start 7 million indemnificatory houses, less than that of 10 million last year but still need large investment. Housing guaranteeing project will finish 5 million houses this year.

广东中行喊停8.5折房贷优惠

近日中国银行广东省分行已经正式发文通知各网点停止首套房贷利率最低8.5折优惠。据中行内部个贷业务人士透露，最低9折优惠执行已有一段时间，下发通知标志着政策正式调整。

BANK OF CHINA, GUANGDONG BRANCH STOPS THE 15% DISCOUNT FOR HOME MORTGAGE

Recently, Bank of China, Guangdong Branch has an official statement noticing every branch to stop the 15% discount for the first home mortgage. According to the credit clerk in Bank of China, 10% discount for the home mortgage has worked for quite a period and the new notice means the formation of an official regulating policy.

房地产投资增速等指标年内首现回升

国家统计局9日发布的数据显示，今年1~8月全国房地产开发投资同比名义增长15.6%，增速比1~7月提高0.2个百分点；8月份房地产开发景气指数为94.64，比7月提高0.07点。值得注意的是，这是房地产投资增速等指标年内首次回升。

THE INVESTMENT INDEX OF REAL ESTATE FIRST CLIMBS BACK WITHIN THE YEAR

Data from State Statistics Bureau shows that from January to August, the development investment on national real estate rise by 15.6% on year—on—year basis, which bears an increasing speed of 0.2% than the last 7 months. Business development index of real estate in August is 94.64, increasing 0.07% than that in July. It is worth noting that this is the first time that the investment index of real estate climbs back within the year.

中国房企拓展海外市场

9月4日，香港信报援引消息称，内地房企碧桂园已计划投资约21亿元人民币在马来西亚的两个项目上，其中双文丹项目预计于明年上半年推售。碧桂园官网披露的信息显示，在马来西亚现有的两个项目——双文丹（Serendah）和土毛月（Semenyih）均位于雪兰莪州，且这两个规划为纯别墅社区的项目，距离马来西亚首都吉隆坡均较近。

CHINESE HOUSING ENTERPRISES EXPAND FOREIGN MARKET

On September 9th, HK newspaper quotes the news that Bi Guiyuan has planned to invest 2.1billion yuan on two projects in Malaysia, one of which Serendah project will be for sale during the half of next year. Its official website said that the two projects, Serendah and Semenyih, both locate in Selangor, not far from Kuala Lumpur and will be pure villa community projects.

金九银十首周54个主要城市成交量环比降17%

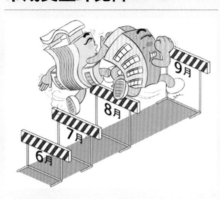

中指院数据显示，9月3日~9日的一周内，虽然仍有八成城市楼市成交同比上升，但增幅明显趋缓。环比方面，在中原地产监测的54个主要城市中，新建住宅签约量下降近17%，其中北京下降19.6%，广州下降21.4%，深圳降幅超过两成。

TRADING VOLUME IN 54 MAIN CITIES DROPS IN A LINK RELATIVE RATIO OF 17% IN THE FIRST WEEK OF SEPTEMBER

Data from CREIS (China Real Estate Index System) said that from September 3rd to 9th, though 80% urban real estate market gain a year—on—year increasing trading volume, the rate of increasing is slowing down. Among the 54 main cities monitored by Centaline Property, the signed number of the new houses drops in a link relative ratio of 17%. Beijing drops by 19.6%, Guangzhou 21.4% and Shenzhen over 20%.

轨交建设获批 助推多城房价上涨

国家发改委公布的信息显示，19个城市25个城市轨道交通项目获批。"房价跟着轨交涨"一直被视为楼市的"黄金法则"，即使面临严厉调控，依然是市场热捧的"香饽饽"。此次相关城市轨道交通建设规划密集通过，再一次给这部分城市的房价带来"想象空间"。中国指数研究院发布的报告认为，除去市场因素，一条轨道交通平均会为沿线房价带来8%~20%的上涨。

RAIL TRANSIT URGES THE RISE OF HOUSE PRICE

Info from NDRC (National Development and Reform Committee) shows that the construction of 25 urban rail transits in 19 cities has got approved. Housing price going up with the rail transit has always been a golden rule in the housing market, active all the time though facing strict regulate and control. At the moment, the intensive approval of urban rail transit construction brings increasing space for the housing price again. China Index Academy made a report to state that without taking the market factor into consideration, one rail transit line can bring the housing price up by 8% to 20%.

万科斥资逾60亿再拿地

9月11日，万科以15.8亿竞得广东顺德德胜商务区一地块，溢价率14.38%，成交楼面价为3 488.66元/平方米，成为顺德新总价地王。仅仅一周前，万科刚刚"豪掷"46.71亿元分别在广州、合肥拿地。据统计，9月开始的这11天，万科已斥资逾60亿在全国拿地；而自7月份起，万科在土地市场上的投入已超过125亿元。

VANKE INVESTS 6 BILLION YUAN AGAIN FOR LANDS

On September 11th, Vanke bought a parcel in Desheng Business Block, Shunde, Guangdong with 1.58 billion yuan. The premium rate of 14.38% and the accommodation value of RMB 3,488.66/m² make Vanke become the new imperial estate in Shunde. Just one week ago, Vanke spent 4.671 billion yuan for lands in Guangzhou and Hefei. According to the statistics, during the first 11 days in September, Vanke has spent 6 billion yuan for lands. From July, the investment in land market has been over 12.5 billion yuan.

Castlecrag住宅

这是由CplusC Architectural Workshop事务所设计的位于澳大利亚悉尼的一个住宅项目，项目将一个住宅进行扩建，将后墙拆除，看起来好似一个娃娃屋。新的黑色木质盖顶创造出一个屋顶和侧面的墙壁，它们围绕首层和二层的房间和阳台，而玻璃墙壁则滑动打开，连接起居，厨房和花园。

Castlecrag House
Australian studio CplusC Architectural Workshop has extended a house in a Sydney suburb so that it looks like a doll's house with the back wall taken off. A new black—stained timber canopy creates a roof and side walls around rooms and terraces on the ground and first floor, while glass walls slide open to connect the living room and kitchen to the garden.

画廊住宅

项目包含画廊及住宅两种功能，由一个线性的建筑及其内部的一个扭曲形体组成。这个扭曲的内部体量向上升起，穿过屋顶，容纳卧室及其他私人空间，它与首层三倍高的画廊空间相分离。一个盒子形的楼梯将人们从画廊引入到上层，之间穿过一个穿孔金属挡屏。

Gallery House
This combined house and gallery comprises a single rectilinear building with another twisted inside it. The twisted inner volume towers up through the roof to house the bedrooms and other private spaces of the residence, keeping them separate from the triple—height gallery on the ground floor. A boxy staircase leads up from the gallery to the floor above and passes by a perforated metal screen that shields a window to the bathroom.

宅上宅

这是由THE Architectes事务所设计的位于法国巴黎的一个住宅扩建项目，它包含一个神秘的黑色尖顶结构，它附着在一所郊区住宅的屋顶，内部包含两个新的卧室空间。屋顶的黑色木结构与住宅的陶砖及白色背景形成对比。尖顶的斜坡与既有的房屋屋顶轮廓相协调。

Une Maison sur la Mainson
A mysterious black gable frames two new bedrooms on the roof of a house in the outskirts of Paris. Designed by French studio THE Architectes, the black—painted timber extension contrasts with the clay tiles and white render of the house's walls and roof. The slope of the gable matches the angle of the existing pitched roof.

尖顶住宅

这是由on3 architekten事务所设计的位于瑞士的一个住宅项目，它有尖尖的末端墙壁，看着好似沿着中心垂直接缝线折叠。一个暴露的混凝土楼梯连接三层空间，看似漂浮在一道金属线条形成的帘子后面。一个巨大的屋顶窗增加了室内的活动空间，这里鸟瞰下面的走廊空间。

Wohnhaus Ginkgo
This concrete house in Switzerland by Basel studio on3 architekten has gabled end walls that appear folded along central vertical seams. An exposed concrete staircase connects the three storeys of Wohnhaus Ginkgo and is suspended behind a balustrade of taught wires between the two upper floors. A large dormer window increases the amount of inhabitable space on the top floor, which overlooks the corridor on the floor below.

H形住宅

这是由57STUDIO事务所设计的位于智利圣地亚哥的一个住宅项目，项目周围最大的特点是古老的树木，空间排布形成一种H形的布局，它与场地完美匹配，保护了树木，也创造了一个突出树木存在感的庭院空间。

Fray León House
The house is located in a district in the east area of Santiago, and one of its characteristics is the presence of old trees. The spaces are organized in an "H" shaped plant adapted to the lot, having great care for the trees and creating patios that reinforce their presence from the inside.

ASH住宅

这是由I.R.A事务所设计的位于日本北海道的一个住宅项目，因为位于寒冷的地带，住宅设计的重点围绕屋顶和雪展开。设计师打算设计一个六十度倾角的屋顶，它可以减少不同程度的积雪积压，同时生成舒适的室内环境。

ASH House
When considering the house built on the heavy snowfall background of the arctic cold which will be midwinter—20℃, the architects paid their attention to "snow" and a "roof", and proposed the house of the roof displeased extremely. By considering it as a 60—degree steep slope roof, various snow damage and snowy load are mitigable only with the natural snow coverage to the ground. The form produced as the result makes comfortable interior space.

Juso护理中心

这是由Saraiva + Asociados事务所设计的位于葡萄牙卡斯凯village的一个护理中心项目,这个项目是对一个既有的电子工厂空间的改造,建筑师的重点是维持原有建造的体量特征。因此最后的设计主要针对室内展开,同时包裹在一个新的外立面表皮之下。

Juso Continuing Care Unit

The Project proposal was to create a Medium to Long—Stay and Convalescence Health Care Unit. The design created by Saraiva + Asociados will be located in an existing building, the Headquarters of "Standard Eléctrica" (Electric Factory), located in Aldeia de Juso. One of the architect's main concerns was maintaining the volumetric characteristics. As a result the intervention reveals mainly the interior spaces alongside a new exterior finish.

瑞尔森大学影像中心

翻新和扩建的建筑外立面是双层的玻璃表皮,它下面隐藏着LED照明系统,夜晚这些照亮的玻璃板面延展三个表皮,它们可以统一点亮,也可以分开点亮,包含大约1 670万种不同的颜色组合。白天这些顶层的不透明玻璃形成白色的流畅背景,与中心下层的透明结构形成对比。

Ryerson Image Centre

The renovated and expanded building's exterior has a double—skin glass cladding that conceals its LED lighting system. At night, illuminated glass panels across three facades glow separately or in unison with a possible 16.7 million different color combinations. By day, this opaque glass surface on the upper floors provides a seamless white backdrop to bustling campus life and contrasts the Centre's transparent glazing at ground level.

广州CTF塔楼

这是由KPF设计的广州CTF塔楼,该项目正在建造过程当中,它位于广州珠江新城CBD。塔楼的形式受到内部项目的影响,形成四个主要过渡部分的向内收缩,塔楼开始于办公,而后住宅,再而酒店,直到最后的顶端。

CTF Guangzhou

KPF recently shared with us their latest design for a 530 meter mixed—use tower in Guangzhou's Zhujiang Xincheng CBD — the CTF Guangzhou — which is currently under construction. The form of the tower is informed by its internal program through setbacks at four primary transition levels; starting with the office to residential, residential to hotel, hotel to crown, and lastly crown to sky.

山丘场馆

这是由Studio 44设计的位于俄罗斯的一个场馆项目,方案将场馆包含在一个绿色的山丘上,山丘中心是足球场,并含有三层的坐席,能够容纳33 000名观众,场馆顶部是一个上层看台,这里能容纳另外9 000人。

Hill Stadium

"Studio 44" designs the football stadium as a green hill in the heart of which lies the football field (105 x 65 m) with a three—level amphitheater with 33 thousand spectator seats, while its top is crowned a snow—white cup of the upper stands capable of receiving yet another 9 thousand people.

康莱德酒店

酒店位于闹市街区附近,周围都是普通的办公大楼,网格子的玻璃幕墙到处都是,而这个酒店大楼则不同,它的表皮旨在活跃建筑所在的场地,与单调的直线相区分,它将直线扭曲,如同人身体上的细胞组织,它会随着延展和环绕在厚度上发生不断的变化。

Conrad Hotel

Although the complex is within close proximity to the lively pedestrian shopping boulevard, Wangfujing Street, the immediate surroundings of the central business district are a conglomeration of mundane office buildings whose exteriors are lined with a grid of curtain wall windows. To contrast the established urban fabric, the facade of the hotel intends to revitalize the area and break away from the monotonous rhythm by warping the orthogonal lines. Similar to nervous tissue, the organic envelope changes in thickness as it is stretched and wrapped around corners.

九世王公共项目

这是由Shma事务所设计的位于泰国曼谷的一个大型公共广场项目,它融合一个学习中心和展示大厅,是为纪念九世王而建。项目位于历史性环境之中,方案呈现山形的地标结构,反映泰国古老的智慧,同时呼应现代社会的活动。

LCPK

Shma has won 1 in 5 placse for LCPK competition, a design for public plaza integrated with learning centre and exhibition hall in honor King Rama IX. Situated in the foremost historical district of Bangkok, our proposed mountain—shaped landmark is the reflection to Thai ancient wisdom while responding to the activity of modern society.

FRENCH STYLE PALACE COMMUNITY: DIGNIFIED, ELEGANT, NATURAL AND HUMANISTIC

| Royal Spring Villa, Beijing

尊贵 典雅 自然 人文的法式宫廷社区 —— 北京金科·王府

项目地点：中国北京市	**Location:** Beijing, China
建筑设计： 筑博设计集团（北京）股份有限公司	**Architectural Design:** Zhubo Design Group (Beijing) Co., Ltd.
用地面积： 152 307 m²	**Land Area:** 152,307 m²
总建筑面积： 186 323 m²	**Total Floor Area:** 186,323 m²
建筑密度： 30%	**Building Density:** 30%
容 积 率： 1	**Plot Ratio:** 1
绿 地 率： 35%	**Greening Ratio:** 35%

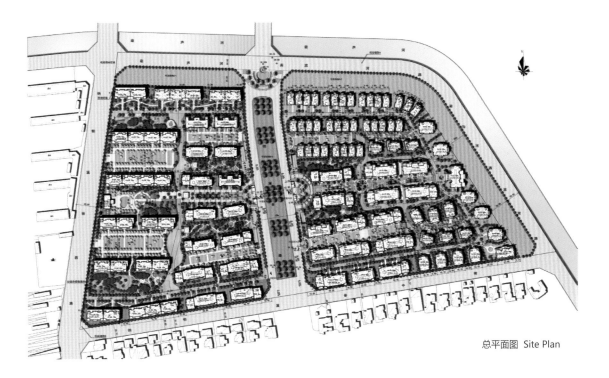

总平面图 Site Plan

MASTER AND MASTERPIECE | 名家名盘

项目概况

金科王府项目位于北京昌平九华山庄向东300 m路南、传统的小汤山温泉度假区核心位置，项目占地15万 m^2，容积率接近1。项目定位为品味、精致的居住、温泉休闲度假社区，打造空间与时间完美结合的豪宅生活。项目东、北两侧毗邻温榆河的支流葫芦河，项目周边遍布成熟居住社区和高档休闲设施，因此，设计师充分利用项目用地的特性，力图创造具有丰富文化和自然景观相结合的法式宫廷社区。

规划布局

项目总平面设计：两个地块在基地朝向纳帕溪谷中路的中间部位各设小区的主要出入口，另在地块南侧各设置一个次要出入口，东侧B—5地块以2~3层的低层户型为主，园区内组团分明，各楼间错落排布，组团间穿插休闲活动场地，滨水设置观景户型，建筑体量较小，与西侧B—2地块的花园洋房形成产品的互补。西侧地块靠近纳帕溪谷西路为7~8层的花园洋房，靠近纳帕溪谷中路一侧为2~3层的联排户型，中间为一条景观水系，靠近水系的户型适当加大，使其景观价值最大。

社区规划将以组团形式呈现，强调组团的私属性与景观均好性，将社区的主入口空间扩大，结合景观与城市外部空间形成有机的过渡。在建筑的形体处理上，靠近水系的洋房层数适当降低，与东侧联排体量上相互呼应，形成较缓和的天际线。

建筑设计

建筑风格塑造：建筑形式为法式宫殿风格，配以法式宫廷园林，多采用对称造型。整体设计注重欧式古典风格比例尺度的推敲，古典建筑构件设计表达清晰，运用不同的古典语汇，营造高贵雅致的建筑氛围。建筑采用浅色全干挂石材外立面，结合葡萄牙米黄的天然石材肌理，色彩既明快又不失稳重，体现了法式宫廷建筑的尊贵与典雅。

建筑平面设计特色：本项目建筑形式多样化，既有强调邻里关系的联排住宅，也有注重私密空间的独栋；既有介于联排与独栋之间的"类独栋"住宅，又有花园洋房系列。在单体平面设计上，依据不同的建筑形式，而做出不同的设计处理。比如：花园洋房通过层层退台的方式形成空中的花园露台；联排住宅的端户平面设计上，采用侧入、侧向私家花园的方式，又赋予其独栋住宅的品质。

无障碍设计：社区内园林均设置无障碍盲道，在西侧B—2地块设置了无障碍户型。

景观设计

东西两地块出入口处，空间开敞，配合景观水系，形成每个园区的集中景观区域；另外每个组团中间也有相对独立的法式景观区域，强调中心感与几何对称；园区内一条水系蜿蜒而过，使这些组团相对独立又互相串联，构成景观主题节点。结合入口处弧形的景观构筑物，形成车行人行入口的标志性景观。

MASTER AND MASTERPIECE | 名家名盘

Profile

Royal Spring Villa locates the core area of Xiaotangshan Hotspring Resort, 300m south of Jiuhua Resort and Convention Center in Changping District of Beijing City. Occupying a land area of 150,000m^2, with a plot ratio reaches 1, it is envisioned to be a high—taste and boutique hotspring resort which provides perfect villa life. Hulu River, the branch of Wenyu River is on the east and north side of this project. There are hundreds of old trees within the site as well as mature residential areas and high—end resort facilities in surrounding areas. Thus the designers make full use of these advantages, trying to create a French style villa community which combines culture with natural landscape perfectly.

Planning and Layout

Site Plan: Main entrance of these two plots open to the middle of Napa Valley Middle Road, and the sub entrances are set on the south side. B—5 plot on the east is mainly designed with 2~3—storey villas which are well arranged in groups with varied entertainment and activity arenas in between. Building density is lower here, complemented to the garden houses in the west B—2 plot. 7~8—storey garden houses are built nearby Napa Valley West Road, and 2~3—storey townhouses are set along Napa Valley Middle road. With a landscape river between these two groups, the products nearby the river are designed with large areas to get the maximum landscape views.

The community is composed of building groups. It highlights the privacy and landscape views of each group. Main entrance squares are expanded to well connect with the urban space. In terms of building height, the garden houses nearby water are designed to be lower, echoing the townhouses on the east and forming a soft skyline.

Architectural Design

Architectural Style: the buildings are designed French palace style together with French—style gardens. All of them are arranged symmetrically. The overall design pays attention to the proportion of classical European style and applies different classical elements to create high—end and elegant atmosphere. Light—colored hanging stones and Portuguese beige stones are used in outer facade, looking bright and dignified to show the royalty and elegance of French style architectures.

MASTER AND MASTERPIECE | 名家名盘

Floor Plan: there are many types of housing products, namely, the townhouses emphasizing neighborhood relationship, the single—family houses paying attention to privacy, the residences similar to townhouses and single—family houses, and the multi—storey garden houses. In terms of the floor plan, different housing products are designed according to their building forms. For example, garden houses are designed in setback style to provide sky gardens and outdoor terraces; townhouses are designed with gardens on two sides to provide great privacy and beautiful views.
Barrier—free Design: the gardens of this community are designed with barrier—free sidewalks and in the west B—2 plot it has especially designed barrier—free units.

Landscape Design

The entrances of two plots are designed with open spaces which, together with landscapes and waterscapes, become the landscape centers of this community. In addition, in every building group, there are independent French—style landscape areas which are designed symmetrically. A zigzag river runs through the site to separate as well as connect these groups. The arc—shaped landscape architecture at the entrance becomes the landmark of this community.

COLORFUL AND BREATH—TAKING VISION, INTEGRAL AND PERFECT LAYOUT | Crape Myrtle • Upper Town, Xi'an

丰富而震撼的视觉 完整而统一的布局
—— 西安紫薇·尚层

项目地点：中国陕西省西安市
规划设计：美国开朴建筑设计顾问（深圳）有限公司
占地面积：58 933.33 m²
建筑面积：180 000 m²
绿 化 率：30%

Location: Location: Xi'an, Shaanxi, China
Planning and Design: CAPA Architecture Design Consul tant(SHENZHEN) Ltd. USA
Land Area: 58,933.33 m²
Floor Area: 180,000 m²
Greening Ratio: 30%

项目概况

紫薇·尚层项目位于西安市太白南路与丈八东路交汇处东北角，北邻百米城市绿化带，未来高新行政办公中心区，紫薇尚层项目占地面积约为58 933.33 m²，是集住宅、商业、LOFT、公寓为一体的多元化、复合性综合社区，紫薇尚层总建筑面积约为180 000 m²，绿化率30%以上，在一个建筑综合体中实现了共存和互补。

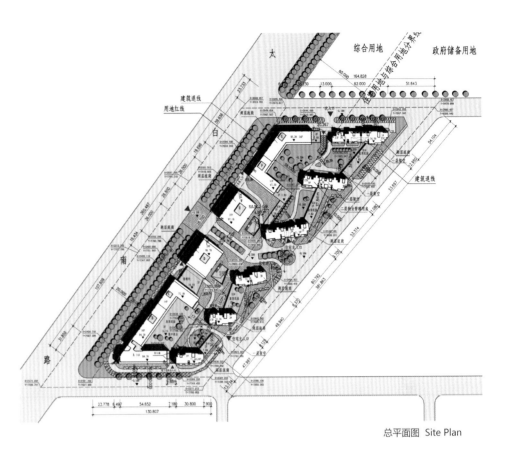

总平面图 Site Plan

商业南区一层平面图
Commercial South District 1st Floor Plan

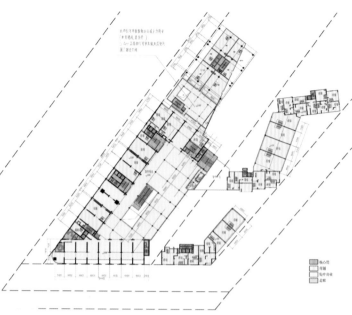

商业南区二层平面图
Commercial South District 2nd Floor Plan

规划布局与建筑设计

规划意向：设计师在社区规划理念上提出在同一地点解决居住，工作和娱乐的生活模式。便于步行、非机动车通行及建立公共交通设施的形态及规模，并具有一定程度的紧缩性以便于人们之间的社会性互动。

规划形态：设计师强调现代构图冲撞、错动、扭转等构成原则，以极具视觉冲击力的点、线、折线构成一系列动态、抽象的构图。贴合地形的不规则形态，强调出城市界面与内部界面的丰富性，形成三维立体的空间特色。

规划构思：

1.针对基地与朝向及城市主干道的交角，方案首先建立了正交与平行道路的两组网格关系，使之成为建筑形体变化的参照系。

2.基地西侧昭示性较强，同时公寓产品具有一定的灵活性，所以西侧立面是体现社区城市形象的主要展示面。

3.两块用地性质的不同在规划中自然形成了一条南北方向的主要轴线，同时在用地的中心位置设计了一条开放的东西轴线，两条轴线建立起规划的基本骨架。

4.规划形态上呈现外部围合、内部开放；周边围合，东侧开放的空间意向。以连续的折板形态与点式、板式建筑形成非对称的动态构图形式。

5.南北两侧形成两组半封闭式的院落空间，与中部流动开敞的开放性空间形成对比。同时在功能上形成互动。

6.产品的分布形成由周边向中心开放型逐渐加强的圈层结构，周边布置私密性最强的住宅，其次为公寓式办公，中心为办公建筑与商业广场。

7.总体空间形态体现视觉的丰富性与震撼力，以完整统一的规划布局体现出空间起承转合的微妙与多重空间的层次。

Profile

Located in the northeast corner of the intersection of Taibai South Road and Zhangba East Road, Xi'an City, the project is near to the hundreds meter long green belt and the future administrative center on the north. Occupying a land area of 58,933.33 m², with a total floor area of 180,000 m² and a greening ratio of 30%, it is a residential complex integrating residence, commerce, LOFT and apartment.

Planning and Architectural Design

Purpose: when making this plan, the designers presented the model of living, working and entertaining in one place. It will includes walking system, non—motor paths and public transit facilities within a compact layout to promote social communications.

Form: the designers emphasize collision, stagger and rotation in modern composition to use amazing points, lines and fold lines to form a series of dynamic and abstract layouts. According to the irregular topography, it highlights the difference between urban interface and internal interface, and then forms a three—dimensional space system.

Concept:

1. With the crossing angle between the site and the urban artery, it first

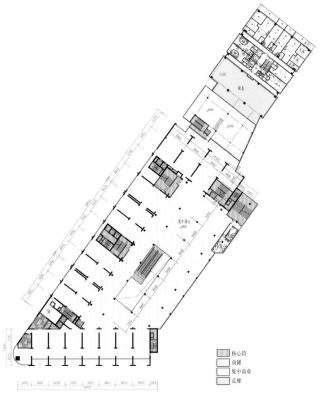

商业南区三层平面图
Commercial South District 3rd Floor Plan

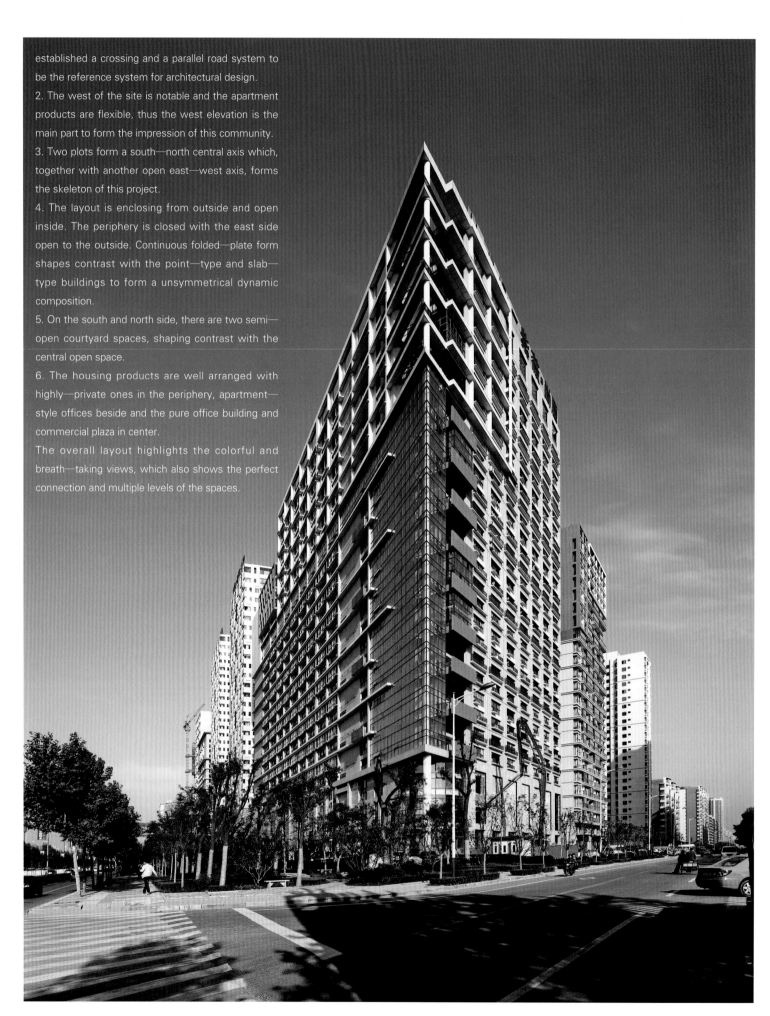

established a crossing and a parallel road system to be the reference system for architectural design.

2. The west of the site is notable and the apartment products are flexible, thus the west elevation is the main part to form the impression of this community.

3. Two plots form a south—north central axis which, together with another open east—west axis, forms the skeleton of this project.

4. The layout is enclosing from outside and open inside. The periphery is closed with the east side open to the outside. Continuous folded—plate form shapes contrast with the point—type and slab—type buildings to form a unsymmetrical dynamic composition.

5. On the south and north side, there are two semi—open courtyard spaces, shaping contrast with the central open space.

6. The housing products are well arranged with highly—private ones in the periphery, apartment—style offices beside and the pure office building and commercial plaza in center.

The overall layout highlights the colorful and breath—taking views, which also shows the perfect connection and multiple levels of the spaces.

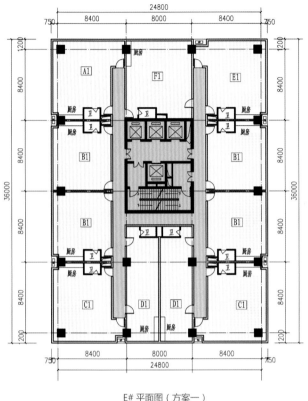
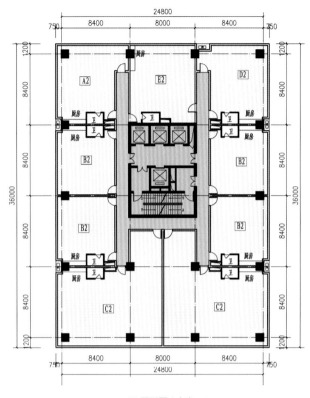

E# 平面图（方案一）
E# Floor Plan (Plan 1)

E# 平面图（方案二）
E# Floor Plan (Plan 2)

BRITISH STYLE FACADE CREATES PURE CULTURAL COMMUNITY

| CSC Palais Grand—Ducal, New Bay City in Shanghai

英伦风格立面营造纯净人文社区 —— 上海新江湾城中建大公馆

项目地点：中国上海市
开 发 商：华东中建地产有限公司
建筑设计：上海天华建筑设计有限公司
总用地面积：114 517 m²
总建筑面积：266 045 m²
容 积 率：1.00
绿 化 率：35%

Location: Shanghai, China
Developer: CSC Land Group
Architectural Design: Tianhua Architecture Design Company Limited
Total Land Area: 114,517 m²
Total Floor Area: 266,045 m²
Plot Ratio: 1.00
Greening Ratio: 35%

项目概况

该项目地处新江湾城中央区域，东至江湾城路，南至殷行路，西至政和路，北至国秀路。基地为方形，地势较平坦。距离五角场核心商圈约3km，配套齐全，生活便利，为一处成熟的居住用地。

规划布局

设计中沿西侧、北侧设置多层建筑，沿东侧、南侧设置低层建筑。沿东侧江湾城路布置主要出入口广场，结合下沉广场设计，引入中轴景观带，一气呵成。

总平面图 Site Plan

MASTER AND MASTERPIECE | 名家名盘

建筑设计

根据现状地形控制点标高，建筑设计中考虑尽量处理好本场地与周围道路场地的衔接关系，减少填挖方量，此场地采用平坡式布置。

建筑立面以英伦风格特征为宗旨，体现现代生活及现代人的价值取向和欣赏品位，以纯净的建筑形象于绿树环抱中营造现代城市绿园。整体对称式立面取得控制地位，其垂直向线条与水平线条极为纯净。立面上颇有节奏的开窗形式及栏杆、屋顶构件等的细部设计既着重于整体又不失细腻。建筑立面处理强调文化与个性，运用光与影来塑造建筑形象，充分利用材料的质感与肌理及虚实体量的变化来造型，突出端庄、高雅的风范，起伏的天际线、丰富的立面线以及景致的比例尺度与环境完美结合，形成了一系列不同尺度、类似形态的空间构成的复合空间序列。

景观设计

规划绿化系统布局采用电线结合的方式，通过景观步行道的串接，将绿化系统贯穿于整个小区住宅群体之间，体现整体的布局概念。并加强空间转折点与视觉焦点处的绿化空间设计。将绿化系统与步行系统相结合。景观系统作为总平面形成的重要构成部分，加强了小区的结构特色，构成高尚的人文气质。

Profile

Located in the central area of the New Bay City, the project reaches to Jiangwencheng Road on the east, Yinxing Road on the south, Zhenghe Road on the west and Guoxiu Road on the north. The site is square and flat with a distance of 3000m from Wujiaochang CBD. With complete supporting facilities and convenient traffic system around, it is really an ideal place for living.

Planning and Layout

According to the planning, multi—storey buildings are set along the west and north side, whereas the low—storey buildings are arranged on the east and south. The main entrance square is nearby Jiangwancheng Road on the east. Sunken plaza and central landscape axis are combined together to present a clear layout.

Architectural Design

According to the controlling points of the site, the design pays attention to the connection between the site and the surrounding roads, trying to reduce earthwork. Thus buildings are arranged in slope style within the site.

The facade is designed in British style to highlight the values and tastes of modern people. Elegant buildings stand together to form a modern city garden with green trees around. Facades are designed in symmetrical style with clear vertical lines and crossing lines. The rhythmic windows, handrails, roof components as well as other details exquisitely designed to accord with the overall style. The facade design emphasizes cultural connotation and characteristics by the skillful use of light and shadow. It also takes advantage of the textures of materials as well as the changes of solids and voids, highlighting the elegant and dignified style. Rolling skyline, colorful facades and the well—proportioned landscape keep harmonious with the surrounding environment, forming a series of sequential spaces of different sizes and similar forms.

Landscape Design

The green system is composed in network style. Walkways are applied to connect all the green areas together, making green penetrate into residential buildings. It pays attention to the green design in turning points and visual focuses. Walking system is combined with the green system which is an import part of the general plan, highlighting the structural characteristics and creating strong cultural atmosphere.

设计要注重实现经济、功能与生态的和谐统一
——访北京寰亚国际建筑设计有限公司董事长 赵士超

■ **人物简介**

赵士超，北京寰亚国际建筑设计有限公司董事长、总建筑师、创意设计总监，拥有十二年的社区设计经验，曾先后供职于长春市建筑设计研究院、澳洲U&A设计国际集团、大地建筑事务所（国际），于2008年创办北京寰亚国际设计有限公司，主张以人为本、经济化、实用化、生态化设计为主题理念。代表作有：北京大运河孔雀城、济南世通莱茵小镇、临沂中国铁建·东来尚城、锡盟明珠广场、呼市东方维也纳等。

■ **公司简介**

北京寰亚国际建筑设计有限公司专门从事公共建筑及居住区项目的建筑方案设计。公司与中国相关建筑设计单位建立了长期稳定的合作关系，与北京中翰国际建筑设计有限责任公司（甲级）签订合作协议，充分发挥各自的技术优势，利用先进的建筑设计理念、丰富的建筑设计经验及国际化的设计视角，开拓中国建筑市场。公司服务范围包括城市规划、建筑设计、室内设计、景观设计、工程/建筑投资估算分析等多个方面，目前已在东北、华北、华东、华南等地区设计了诸多工程项目，得到了中国同行与业主的好评及认可。

《新楼盘》：提到行业的发展，就不得不说合作，你们与国内外的很多公司都有合作关系，这种合作是基于一种怎样的平台呢？它对于一个大型项目来说有何重要意义？

赵士超：谈到发展，不得不提到世界经济发展的客观趋势——经济全球化，世界各国不可能离开这一进程而孤立发展，在全球化的国际经济关系中，没有以知识的生产、交换和消费为基础的知识经济和世界经济的信息化，就没有当今的经济全球化。对于建筑行业而言，我们更应该清楚的看待经济全球化的机遇与挑战。为此我更提倡将"合作"看成是一种信息共享性、沟通协调性的平台。在此平台上结合国内外设计机构其他方面的优势，取长补短，从而实现多层面的深入合作；同时又可以促进公司的多元化发展，感受建筑的最新动态，走在国家乃至国际的建筑前沿。

其实这个平台更像是一个大型思想融汇之地，侧重共享与沟通。对于一个大型项目来说，我们需要一系列的过程把握，需要前期的调研，深入的讨论，集思广益来完成前期概念方案，细节的统一……而这所有的一切都需要以这个平台为基础，让我们更有思想，更有效率，更能统筹全局。

《新楼盘》：建筑（规划）设计是一个循序渐进的过程，你们是怎样去处理好各个设计流程之间的关系的？

赵士超：基于对项目的不断完善，我对设计的理解，大概分为七个流程或步骤。第一步是要对所做项目进行全面分析，包括对整个项目的使用功能和使用目的的分析，以及项目最终意图表达的分析；第二步是对项目现场和地形的分析，基于对此的理解，可以使我们做出一个更好的项目设计方案；第三步是对项目周边环境和当地文化的理解；第四步是对项目最终使用者的分析和理解，在这里应该思考的是，如何创建一个更好的且能够满足使用者需求的空间，这才是关键；第五步是基于前述的一系列分析和对项目当地历史性的考察，从而思考项目在这个环境中如何体现历史文脉，寻找建筑本身的文化属性；第六步功能与形式的结合，空间的创造；第七步就是考虑项目与整个自然环境的协调与融入的问题。

除此之外，我们认为，在建筑设计中，艺术性、功能性和技术性都是实现完美建筑不可或缺的元素，针对这些元素，我们也有非常完美的设计参数来支撑建筑设计，可以从技术方面，更好地利用现代科技来促成创意的实现。如今的很多建筑项目都要考虑国际化的因素和当地环境的因素。所以，我们在参与这些项目设计的时候，在项目的概念阶段，一定会作大量的背景调查和环境研究，同时，在建筑的体形上做很多不同的方案。直到最后，引入现代科技的先进手段，才能保证艺术的造型和新颖的设计，最终使完美建筑落地。只有拥有了现代的技术，才能有现代的个性造型和创意，并使其可持续地存在于城市的大环境当中。

《新楼盘》：建筑不是简单地建造房子，而是需要兼顾考虑多种因素，这其中最主要的因素是什么？

赵士超：主要考虑的事情还是自己带领的团队在建筑设计方面的调整和发展。我正在考虑怎样将客户的需求同建筑设计的艺

性、技术性、新材料的应用及工程的现实情况完美地结合到一起。尤其是对我和我的设计团队来说，这是我们最终的设计目标。这个过程是一个循序渐进的过程，而不是突变的过程。为此，我们制定了一个5年计划，希望自己和其设计团队能够在设计的过程中，在不断参与项目的过程中得到一个整体水平的提升。从整个团队规划设计的流程来看，包括分析的过程，设计的过程，监督建造的过程，我们都在不断修正、完善，努力实现自身的设计目标。

《新楼盘》：与大家分享一下近期完成的项目，谈谈它们在设计上有何特色？

赵士超：受呼和浩特名诚地产开发有限公司委托，我公司开始进行"东方维也纳"项目的整体规划设计，本项目位于呼市如意区，是呼和浩特规划总部基地所在地，周边为政府单位及商业用地，同时也是周边为数不多的一块住宅用地，所以在设计时利用本项目独有的地理位置，贯彻寰亚国际以人文本、经济化、实用化、生态化的设计思想，通过对住宅立面及平面功能的精致化处理，在呼市树立了良好的口碑。

"中国铁建·东来尚城"项目，位于山东省临沂市，项目紧临临沂母亲河沂河，具有良好的景观优势，设计时为了满足户户均能享景，在地块北侧设计高层住宅，在地块南侧依靠景观园林，设计花园式洋房住宅，体现出产品的多样性，东来尚城凭借中国铁建的号召力和寰亚国际设计的前瞻性，获得多项奖项。

《新楼盘》：如今很多建筑在注重外部形象的同时也开始关注建筑的文化内涵，应该如何将中国元素合理应用于现代建筑之中？

赵士超：当前，我国设计行业的发展非常迅速，已经成为全球的焦点。世界各国很多的建筑师走进中国，做出了风格迥异的建筑。这种百花齐放的现象出现在中国，既有积极的作用，也有消极的影响。我认为，中国最终将会成为世界经济和政治的领先者，在设计领域，也定会吸引来自世界各地的资源和有能力的设计人才。所以要将国外的专长与国内的特点很好地结合在一起，形成属于自己的特色。

近几年，中国开始注重传统文化的延续，而在设计领域，这一思想也迅速膨胀成一种潮流。除了一些根深蒂固的思想，设计师、策划者们也有意识的加入中国风，以此来吸引大众的眼球。传统元素开始在各个领域广泛的应用，生活中的方方面面都有传统元素的存在。我们必须在对传统艺术表现方式的理解基础上，对传统的元素加以改造提炼和运用，以现代审美的全新视点去重新审视传统文化，把中国传统元素的利用达到最优化，使其更富有时代的特色。要在充分理解传统文化的基础上延其"意"传其"神"，让传统文化在现代标志设计中得到提高和升华，创造出现代、简洁、舒适而又能体现中国精神和意境的现代建筑。

《新楼盘》：居住建筑和公共建筑是你们设计主要建筑类型，这两种建筑形态在设计上面有何不同之处？未来建筑设计领域又将会有怎样的变化与发展？

赵士超：居住建筑从1996年至今在城镇住宅中迅猛发展，年建造量由近3亿平米增至近6亿平米人均建筑面积由近15平米增至约28平米。住宅类型从集合单元式发展为独栋住宅、联排住宅、叠拼住宅等，还出现了情景洋房、花园洋房、空中别墅等。住宅层数从低层到多层、中高层、高层以至超高层。但从总体上来说居住建筑体块相对于公共建筑比较简单（板式住宅和点式住宅），而窗洞的开启位置也相对固定。对于公共建筑来说为了扩大建筑空间的自由度和开放程度，采用了通用的可变空间网架结构，满足了不同功能的使用需求。公共建筑因其规模、体量、高度、功能等等的因素构成了建筑的形象可变与特殊性，也因此有些公共建筑成了城市的标志。研究城市设计问题大型公共建筑和建筑群体对改变城市面貌起着关键性作用，因此公共建筑更侧重从城市整体出发，研究公共建筑及群体对其所属的空间、环境以及相邻建筑的协调关系，做到建筑与城市、建筑与建筑和谐相融。

经历了古典、现代、后现代、新世纪以及世纪末等不同设计思潮的转换，这些"主义"犹如时尚风向标，在很长时间里主宰了设计者与使用者的行为、审美、空间印象以及心理状态。建筑作为人类的基本生产、生活资料之一，其发展趋势是随着社会、环境、经济与科技的变化而不断发展的，并且具有同步一致性的特征。从大众心理变化与社会发展趋势的角度出发，认为未来建筑设计发展将呈现以下几种特点：

风格多样化——信息化社会的到来，社会的更加民主化，物质财富的极大丰富，加之受当代流行艺术思潮的影响，对于不同职业、不同文化背景、不同年龄层次的人来说，有着不同的生活习惯，兴趣爱好，思维方式和价值取向，建筑风格的多样化必将成为当今和未来发展的方向。

绿色环保概念——随着人类对地球环境资源的无度索取导致了许多区域环境的生态系统失衡，人类生存的环境日益恶化，这直接威胁着人类自身的生活质量和生存环境。正是在这形势下全球范围内的环境保护运动蓬勃发展，人与自然和谐共处的"绿色革命"风靡全球，"生态城市"、"生态建筑"、"生态家居"也应时代潮流而被提了出来。

高科技的应用——随着科技前所未有的高速发展和计算机网络技术的广泛应用，人类依托于建筑之上、对于生活与工作的许多梦想和蓝图现已逐渐变成了现实。

Design Should Focus on the Harmony and Unification of Economy, Function and Ecology

—— interview of Zhao Shichao, chief executive of Beijing Huanya International Architecture Design Co., Ltd

Profile:

Zhao Shichao, as the chief executive, chief architect and creative design director of Beijing Huanya International Architecture Design Co., Ltd, has more than 20 years of experience in community design. He once worked in Changchun Architecture Design and Research Institute, Australian U&A International Design Group, Dadi International Architectural Firm and established Beijing Huanya International Architecture Design Co., Ltd, focusing on concepts of people oriented, economical, practical, ecological design. His masterpieces include Beijing Grand Canal Peacock City, Jinan Shitong Rhine Town, Linyi China Railway Construction—East Fashion City, Ximeng Pearl Plaza and Oriental Vienna in Hohhot, etc.

Company Profile:

Beijing Huanya specializes in architectural designs of public facilities and residential projects. It has established a long and stable cooperation with relevant architectural design organizations and signed a cooperation agreement with Beijing Zhonghan International Architecture Design Co., Ltd to fully express respective technical advantages and utilize advanced concepts, rich architecture design experience, international design perspective to explore Chinese building market. The company's service scope includes urban planning, architecture design, interior design, engineering/building investment estimation and analysis which spread widely in northeast China, north China, east China and south China and have won good reputation and acception of Chinese counterparts and owners.

New House: Referring to industrial development, cooperation should be mentioned in any way. As we know, you maintain a cooperation relation with many companies both domestic and abroad. What kind of platform is this cooperation based on? What significance it has for a large project?

Zhao Shichao: When it comes to development, we have to mention the objective trend of world economic development—economic globalization, which every country in the world can't realize development without. In international economic relations amid globalization, if there are no knowledge economy and informatization of world economy which are based on the production, exchange and consumption of knowledge, there isn't the present economic globalization without doubt. Within the architectural industry, we should regard the opportunities and challenges existing in economic globalization. For this, I would like to promote cooperation as a way of sharing of information and coordinating platform. On this platform, we have the opportunity to learn from the advantages of foreign design institutions to make up for our short comings, realize deep cooperation on different levels, boost the company's diversification development, feel the latest trends of architectures and lead the architectural advantage of the country or even the whole world.

In fact, this platform is more like a grand gathering of thoughts, which lays emphasis on sharing and communication. For a large project, we need a series of process control, early—stage investigation, deep discussion and putting heads together to finish conceptual planning and unification of details in early stage. All of these should be based on the large platform to bring out more thoughts, higher efficiency and overall planning.

New House: Architecture planning is a process which should be completed step by step. So how do you deal with the relations of different design stages?

Zhao Shichao: Based on the gradual perfection of projects, my understanding of design could be divided into seven steps. The first step is to conduct overall analysis to the project, including functions and purposes and analysis of final intention expression. The second step is to analyze the site and terrain of the project, which could help us work out a better design scheme. The third step is to understand the surrounding environment and local culture. The fourth step is to analyze the final users of the project. The key point lies in how to create a space in which users' need could be better satisfied. The fifth step is a series of analysis and investigation of local and historical elements, which could help the architecture better reflect historical context and seek for its cultural property. The sixth step is to combine function with form and create the space. The seventh step or the last step is to consider the problem of coordination and integration of the project with the surrounding environment.

New House: Architecture is not simply building houses, which needs to take into account many factors. Which do you think is the key factor then?

Zhao Shichao: The major things to think about are how to achieve team coordination and development in architecture design. I'm thinking about how to perfectly integrate customers' need with the artistry, technicality of architecture design, application of new materials and current situation of the project, which is a final goal for our team. This will be and should be a step—by—step process, instead of a mutation process. Therefore, we have drawn out a five—year plan, which would help myself and my design team improve overall design level in the process of participating in projects. Seeing from the entire flow path of planning and design, which includes analysis, design and supervision over construction, we have been non—stoppingly adjust and improve to realize our goal of design.

New House: Would you please share with our readers your lately finished projects on their special features?
Zhao Shichao: Entrusted by Hohhot Mingcheng Real Estate Development Co., Ltd, we started the overall planning for project "Oriental Vienna". The project is located on the site of Hohhot Planning Headquarter Campus, in Ruyi District, Hohhot, with surroundings being government agencies and commercial land. As one of the few residential plots, the distinctive geographic location is utilized to implement the people oriented, economized, practical and ecological design concept of Huanya International. Through detailed treatment to residential elevation and plane functions, the project has won public praise of Hohhot.
The project "China Railway Construction—East Fashion City" is located in Linyi, Shandong Province, closely next to its mother river—Yi River, possessing find landscape advantage. In order to provide each residence a wonderful view, high—rise buildings were designed at the north end of the site while garden style houses were placed at the south end of the site, neighboring to the landscape garden, reflecting the diversified style of the project. With the special appeal of China Railway Construction and the foresight of Huanya International, East Fashion City has won a series of different awards by far.

New House: At present, more and more architects have started to pay attention to the cultural connotation of architecture, compared to the constantly concerned outside image. So, how do you think architects should apply Chinese elements into modern architecture design?
Zhao Shichao: Today, Chinese design industry maintains a rapid development speed, which has drawn the attention of the whole world. Meanwhile, lots of foreign architects have entered China and demonstrated distinctive architecture styles. This blossoming phenomenon has both the bright side and the dark side. In my opinion, China will eventually become the forerunner of world economy and on the politics arena and she will also attract talents and resources from every part of the world in design field. Therefore, it turned out necessary to combine overseas specialty and domestic properties and form Chinese own features.
In recent years, China has become more and more aware in continuing traditional culture and in design field, it has soon evolved into a trend. Except for some deeply rooted ideas, architects and planners are conscious to add Chinese style into their works to attract the public's eyeballs. Traditional elements have come into play in every different fields and every aspect of life. But it is necessary for us to understand the expression of traditional art first and then transform, refine and apply and appreciate them in modern aesthetic view and achieve the best of application of Chinese traditional elements with modern features. Architects and planners have to extend its meaning and deliver its spirit on condition of understanding the traditional culture and improve and sublimate it in modern landmark architectures, creating modern, simple and comfortable modern architectures which embody Chinese spirit and artistic conception.

New House: Residential and public buildings are two major areas of your designs. What are the main differences of these two types and what changes and developments will they present in future architecture design areas?
Zhao Shichao: Since 1996, residential building has developed rapidly, with annual floor area from 300 million square meters to 600 million square meters while per capita floor area from 15 square meters to 28 square meters. Residential type also evolved from collective units to single—family residence, row house residence and laminated residence, or even scene houses, garden houses and penthouses. Residential layers also increase gradually from low residence to multi layers, medium high—rise residence, high—rise and super high—rise residence. Generally speaking, residential building is relatively simpler (plate type and point block apartment) compared to public buildings, especially with quite fixed position of windowing. However, in order to expand freedom of the space, public buildings apply universal variable spatial grid structure to meet different functional needs. Elements like scale, volume, height, function, etc make up variable image and particularities of architectures, some of which therefore become the landmark of city. In research of urban design, large scale public buildings and clusters play key role in shaping the appearance of cities, which result in the stress of public buildings on city as a whole to analyze their relation with space, environment and buildings in the neighborhood and achieve the harmonious coalesce between architectures and city, architectures and architectures.
Architectures have undergone the transitions of different ideological trends of design, namely classism, modernism, post—modernism, new century and end of the century, which are just like weathervane of fashion, dominating the behaviors, aesthetic appreciation, spatial impression and psychological status of designers and users in quite a long time. As basic production and life material, architecture trends develop with social, environmental, economic and technical changes, with synchronized and consistent features. Starting from public psychological changes and social development tendency, future architecture development would present following characteristics:
Diversified style—coming of information age enables a more democratic society and extremely abundant material wealth. Besides, affected by modern popular artistic trends, people of different careers, background and ages enjoy different living habits, interests and hobbits, thinking modes and values, which invariably result in the diversified architecture styles of modern times and future development.
Green and environmental protection concept—the excessive exploitation of humans to natural environment and resources has lead to unbalance of ecological system in many areas and the gradual degradation of human living environment, which directly threatens living quality of mankind and surviving environment. Just under this condition, environmental protection activities through out the world flourish rapidly while green revolution promoting harmonious co—existence of human and nature blows over the globe and ecological city, ecological architecture and ecological home furnishing are proposed as well.

High—tech application— with the unprecedented development of science and technology and widespread use of internet, many dreams and blueprints of humans which relies on architectures, life and work have gradually come true.

THE INTERGROWTH OF HAKKA CULTURE AND MEDICINE PRESERVATION

| Planning of Nanshufeng King of Medicine
| Health Preservation Valley

客家文化与药物养生的相融共生——南树峰药王养生谷规划设计

项目地点：中国广东省梅州市
开 发 商：广东杉维生物医药集团
景观设计：广州市四季园林设计工程有限公司
占地面积：285 347 m²
建筑面积：28 350 m²

Location: Meizhou, Guangdong Province, China
Developer: Guangdong Shanwei Biomedical Group
Landscape Design: Guang Zhou Shi Si Ji Yuan Lin Design Engineering Co., Ltd
Land Area: 285,347 m²
Floor Area: 28,350 m²

南树峰药王养生谷位于梅州市梅县松口镇，项目占地面积285 347 m²，现状有基本道路和部分生活设施，现状地形地貌，呈"簸箕"状，两则为山脊，中间为山谷，亦可称为"太师椅"状地形，属中国传统文化中认可的招财纳宝、防御自卫、君临天下般的风水宝地。

在规划设计上，设计师把客家文化和药物养生两大理念融入到山水格局中，寓教与游、寓学与游。让前来旅游的非客家人感受客家人的文化传承和产业园的南药文化，以观赏和参与的方式了解传统农耕文化和中国博大精深的中医药文化。

瀑布药田观赏区：原有高山平台形状，宽约100 m，高约25 m，拟设计为假山瀑布和台地式药田结合的方式。斜坡面正面设计宽约30m的帘状瀑布；侧面结合台地式药田设计丝状飞瀑。整体大斜坡以塑石为主体，成为园区内的视觉焦点。

运动健身区：利用原有竹廊"张园"平台，设计亚热带风情露天大泳池、水吧、儿童戏水池等，形成消暑游乐健身的好去处；竹廊南侧设计药疗房作为泳池的配套，散点式以减少对植被的破坏；山凹处聚集溪水并设计为无极山景泳池，体验原生态山林溪水的怀抱。

山顶汤泉游览区：利用现状高山天池，设计为药疗SPA区，纯净的天池水与正宗的南药融合形成多类型

总平面图 Site Plan

NEW LANDSCAPE | 新景观

的药物池，如：当归池、何首乌池等；并设计更衣室、电瓶车场以方便游览浸浴，高山远眺、身心放飞、强身健体。

综合广场售票处：局部回填平整至194m，使之与现状入口广场标高平齐，以便于景观建设，并形成开阔大气的入口区。

酒店生活区：基本位于山脊，以建筑为主体，有多种形态的建筑及多类型的林间游览空间，包括会所式酒店、鸟巢式客房、别墅式客房、养生别墅等。

药王谷水景观观赏区：以整理山谷水系标高为主要内容，提高水的利用价值，使水系统形成五级跌落。五级自然跌落划分为水车参与区、溪涧亲水区、钓鱼区、观水区等不同区域，并最终与客家村落水塘及瀑布连为一体，形成完整的山水格局。

Nanshufeng King of Medicine Health Preservation Valley is located in Songkou Town, Mei County, Meizhou City, covering an area of 285,347 m², with existing roads and living facilitie. The present terrain takes the shape of dustpan or palace chair with the valley in middle and ridges on two sides. This terrain belongs to one of the geomantic treasure land according to Chinese traditional culture, which attracts treasure and possesses the advantage of defending itself and a temperament of emperor.

In planning design, architects mix the two concepts of Hakka Culture and Medicine Preservation into the landscape pattern and teach visitors through tourism and enable visitors to learn through tourism. The coming visitors would experience the Hakka Cultural heritage and the South Medicine Culture in the industrial park and understand the traditional farming culture and Chinese profound traditional medicine culture through appreciation and participation.

Waterfall medicine field appreciation area: the existing terrain displays the high platform with width 100 meters and height 25 meters, which is planned to combine rockwork and waterfall as well as platform—type medicine field. The front side of the slope is designed to be 30—meter wide curtain—type waterfall while the side faces are designed into filiform waterfall in combination of the platform—type medicine field. The entire slope emphasizes on stone—shaping, which stays the focus in the park.

Health and fitness area: utilizing the existing bamboo corridor—Zhangyuan platform to design subtropical outdoor super pool, leisure bar and children's waterpark, etc. for avoiding of summer heat and excising and recreation. Medication rooms are designed at the south side of the bamboo corridor as a supporting facility of the pool in a scattered way to reduce damage to vegetation. The corrie gathers stream and is designed into Wuji Mountainous Landscape Pool for visitors to embrace the original and ecological mountains and forests.

Mountain top hot spring sightseeing area: the existing Tianchi on the high mountain is designed into SPA area, with the mix of pure Tianchi water and classic south medicine to form a multi—type medicine pool, like angelica pool, Polygonum multiflorum pool and etc. There are also changing rooms, electric car parking lot for the convenience of sightseeing and bathing. This place is a great resort of mountain overlook, relaxing body and heart and body building.

Comprehensive square ticket office: the square is partially filled and leveled up to 194 meters high, he same height with the existing entrance for easier construction of landscape and forming of an open and ambient entrance.

Hotel residential area: the residential area is basically located on the ridge, focusing on different forms of architectures and multi—type tourist space in the forest, including club hotel, nest type guest room, villa type guest room and health preservation villa, etc.

King of Medicine Valley water landscape sightseeing area: Settling elevation of valley water system is targeted as the main content to improve the value of water and create a five—step fall, namely waterwheel participation area, mountain stream intimacy area, fishing area and water appreciation area. These areas finally join together with Hakka villages water ponds and waterfalls to form a complete mountain and water pattern.

A ROMANTIC AND EXOTIC SPANISH CULTURAL TOUR

| Vanke Town Landscape in Zhongshan

充满西班牙异域情调的浪漫文化之旅—— 中山万科城市风景

项目地点：中国广东省中山市
开 发 商：中山市万科房地产有限公司
景观设计：深圳市雅蓝图景观工程设计有限公司

Location: Zhongshan, Guangdong Province, China
Developer: Zhongshan Vanke Real Estate Co., Ltd.
Landscape Design: Shenzhen Yalantu Landscape Engineering Design Co., Ltd.

项目位于中山市南区中心区，与石岐区、东区相近，周边公共服务设施配套齐全，交通极其便利。此次设计为整个项目的第五至第十组团及体育公园和公共景观轴。项目建筑为小高层住宅，现代西班牙建筑风格。设计基于建筑的风格，采用现代设计手法，强调直线清晰、简洁明了的线形美感，与建筑外立面线形相统一，同时又有各自的特色。

在设计理念上，设计师试图通过对自然美景的重塑，再现自然界变化万千的景物，以达到不仅在视觉上给人以美感，而且在空间上给人身临其境的环境，让居者寻找到最终的归属与满足。在追求自然、和谐的过程中，无论在哪个角度，人们都可能不经意间发现设计师独具的匠心，从而感受到设计师对西班牙异域情调的浪漫文化之旅和对悠闲生活方式体验的倡导。

在设计细节上，追求粗犷、手工化，大量运用了手工砌抹的墙面、精心锻造的铸铁艺术、质感自然亲切的陶土砖、制作精美的手工绘制的彩色瓷片，从而体现韵味浓厚的西班牙异域情调，体现艺术和浪漫的生活方式。通过多种层次、多种质感的植物搭配，营造丰富而又充满情趣的空间。在铺装材料上，主要应用于非重点区域的局部园路铺装，体现其自然亲切的质感，规格以散板为主，铺设较为方便，色彩变化丰富，富有自然休闲的温馨浪漫的气息。

设计以点带线，以线带面，注重景观的整体性；同时以绿化为重点，加强植物对景观效果及空间的营造。项目创造了一个统一而不单调，丰富而不凌乱的空间；同时，注重休闲活动与游乐功能，做足了人性化设计，营造出一个适合居住的生活空间。不管以什么题目作为设计主旨，使附近居民都能寻找到适合休憩和使人们舒缓平和的空间氛围，感受到浪漫的文化之旅和对悠闲生活方式的体验，这些既是设计的根本前提，亦为目标。

在植物种植上，以树型优美、高大挺拔的景观大乔木为主景，结合浓密的绿化背景和开阔的缓坡，营造开阔疏朗的景观效果，并在其中点缀彩色叶植物，突出植物的季相变化。加上简洁舒适的铺装、高大的耸立的乔木，勾画出一幅安静惬意的画面，营造一种祥和安稳的居住环境，同时创造更多灵活性的空间，使人们获得更多的体验方式。

NEW LANDSCAPE | 新景观

Located in the centre of south district in Zhongshan City, the building as the fifth and tenth group and a sports park as well as public landscape axis is adjacent to Shiqi district and east district. And the public service facilities in surroundings are complete, so the traffic is very convenient. The residential buildings in this area are middle height and with modern Spanish architectural style. On the basis of architectural style, modern design techniques are used to highlight the distinctness of straight lines and the linear aesthetics of simple and clear, thus consistent to the exterior lines of building's exterior facade with their own characteristics.

In design concept, architects tried to restore natural beauty and reproduce the ever-changing landscape of nature to achieve aesthetics visually on one hand, and present residents with an immersive environment physically on the other hand. In this way, residents could discover their final sense of belonging and satisfaction. In pursuit of nature and harmony, people could find architects' unique craftsmanship unintentionally from every different angle, which help them to experience the promotion of architects for romantic culture and leisure lifestyle of Spanish temperament.

In detail design, architects chose rough and manual way by utilizing a large amount of manual laid bricks, elaborately wrought iron cast art, terracotta bricks of natural texture and colorful tiles of exquisite fabrication in order to embody intense and classic Spanish tone and an artistic and romantic life style. Through collocation of multi-layer plants of different textures, a rich space full of interests was created. When choosing fitment style, architects applied partial garden road style for non-key areas to demonstrate its natural and amiable texture. Loose plates were utilized widely for its convenience of laying, abundant changes of colors and natural and romantic sense.

Architects' design utilized points to drive lines and lines to drive surface and highlighted the integrity of landscape. Meanwhile, with greening as the key part, architects also strived to use plants in forging good landscape effect and space creation. Finally, the project was created into a space unified but not dull, lavish but not messy. At the same time, leisure activities and amusement functions were both stressed to create more humane and comfortable living space. Whatever subject was chosen as the theme and purpose of design, the fundamental premises were always similar: enable the residents to find a spatial environment which is more appropriate and soothing for taking a rest, to experience a romantic cultural journey and a leisure and easy lifestyle.

When choosing the plant, designers regard graceful, tall and straight megaphanerophyte as main feature, combine bushy greening background and open gentle slope to create open and clear visual effect, add plant with colorful leaves to the landscape highlights season change of the plant. Concise and comfortable pavement and tall megaphanerophyte draw a peaceful and cozy picture, create a peaceful and steady living environment, meanwhile, more flexible spaces have been created, make people acquire better experiences.

FEATURE | 专题

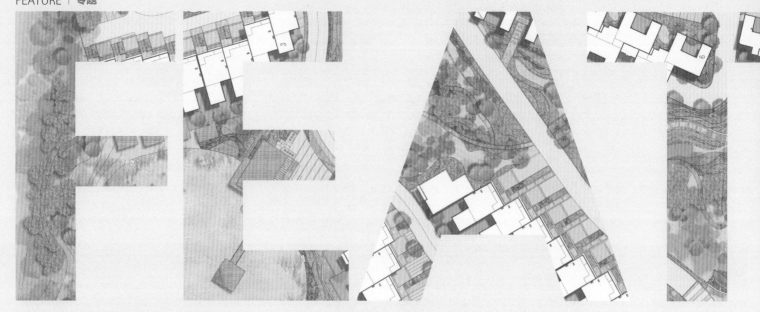

总部科技园

专题导语

作为总部经济的主要载体，总部科技园区是总部经济和商务园结合的新事物，是一种新兴的园区规划和建设模式。随着总部经济的发展，我国的总部科技园区建设快速发展起来，由于我国的总部科技园区建设尚处于起步阶段，因此该类项目的规划和建设实践中仍面临诸多问题。具体而言，总部科技园项目该如何定位，其区位及功能组成如何，怎样建设符合园区开发模式、发展方向的单体建筑等问题都亟待设计师们去解决。本期专题精心挑选多个国内外优秀总部科技园项目，旨在为广大的设计师及设计爱好者们提供参考借鉴。

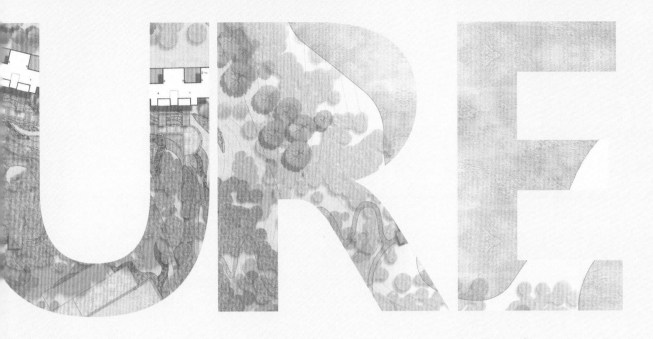

Introduction

As the carrier of headquarters economy, headquarters sci—tech park is the new product due to the combination of headquarters economy and business park. It is a new model for park planing and construction. With the development of headquarters economy, the construction of headquarters sci—tech parks speeds up. And at the present stage, the planning and construction of this kind of project faces many challenges. How to define its orientation, how to arrange its functions and how to design each single building, all are problems need to be solved. In this issue, we have selected many typical headquarters projects both at home and abroad, to be great reference to the architects and design lovers.

北京中视东升文化传媒有限公司

企业概况

北京中视东升文化传媒有限公司是一家立足于文化传媒事业发展的综合性企业。2008年以来，公司与朝阳区高碑店高井村合作，投资开发和运营北京东亿国际传媒产业园，已成为专业化的文化产业园区开发运营商、文化产业服务提供商和文化产业投资参控股商。中视东升旗下拥有北京东亿天弘影视文化有限公司、新维亚商务酒店管理有限公司、北京东亿方元物业管理有限公司等多家全资和控股子公司。

产品与服务

2010年1月，中视东升投资建设的7万 m^2 一期项目已竣工并投入使用，成功构建了以高清数字演播大厅为主体的影视制作基地和与中国传媒大学合作的创新人才培养基地；2011年8月，二期项目共14万多平方米已开工建设，即将构建传媒企业总部、文化创意孵化两大基地。

中视东升通过对资源精心布局、产品差异化定位，打造影视制作、创新人才培养、传媒企业总部、文化创意孵化四大产业基地，将把园区建设成为在北京乃至全国具有影响力的高品质、国际化、生态型的产业聚集区和孵化器，成为首都文化创意产业的新地标。

ZHONGSHIDONGSHENG CULTURE & MEDIA CO., LTD

Company Profile

Beijing ZhongShiDongSheng Culture & Media Co., Ltd is a integrated enterprise focusing on the culture & media business development. Couple with Gaojing Village (in Gaobeidian, Chaoyang District), the company has become the developing operator, cultural business service provider and cultural business investment controlling supplier of the industry park with professional skills by investing and operating Beijing Dongyi International Media Park . It owns or partly owns several subsidiaries including: Beijing DongYiTianHong Film and Entertainment Co., Ltd., New Via Hotel Management Co., Ltd. and Beijing DongYiFangYuan Property Management Ltd.

Product and Service

In January 2010, the company invested a 70,000 m^2 project, project I has been completed and put into use, successfully built a high—definition digital performance hall as the main film and television production base and an innovative talent training base in cooperation with China Media University; In August 2011, project II (some 140,000 m^2) has started construction and will soon become the media corporate headquarter and the cultural innovation incubator bases. Through careful layout of the resources, product differentiation, positioning, to create a film and television production, innovation and personnel training, media, corporate headquarters, cultural and creative incubators four major industrial bases, Dongsheng will built the park as an influential, high quality, international, eco—industrial and innovative place of Beijing or even of the whole Country and as a new landmark of the capital's cultural and creative industries.

建筑：人与人之间的情感交流平台
——访北京博地澜屋建筑规划设计有限公司总设计师　曹一勇

■ 人物简介

曹一勇
世界华人建筑师协会资深会员
亚太商业不动产学院学术委员会委员
全国工商联商业不动产专委会专家委员
北京博地澜屋建筑规划设计有限公司总设计师
国家一级注册建筑师、高级工程师、建筑学硕士

■ 工作经历

1990.7—2005.7 邮电部设计院（后更名为：中询邮电咨询设计院）总建筑师
2005.8—2010.1 五合国际（WERKHART）建筑设计集团
集团下属北京子公司总经理
集团下属北京华特建筑设计顾问有限公司董事、总经理
2010.1—至今 北京博地澜屋（BUILDINGLIFE）建筑规划设计有限公司 总设计师

■ 主要作品

汕头粤东信息大厦、四川移动总部办公大厦、中国电信重庆第二长途通信枢纽楼、中国电信拉萨长途通信枢纽楼、中国电信贵阳第二长途通信枢纽楼、中国电信成都第二长途通信枢纽楼、中国电信上海信息产业园、新华书店南京物流中心、北京东亿国际传媒产业园、承德博物馆、河南工业大学新校区、山东烟台芝罘岛总体规划、北京春天MALL规划、承德御道风情商业街、卓达威海香水海四期、承德双滦人民医院、天津环秀湖旅游度假别墅区等。

《新楼盘》：将规划、建筑、景观、室内四大专业的融合渗透是你们设计的独特优势，这种多目标的整合有何重要的意义？

曹一勇：将规划、建筑、景观、室内四大专业统筹考虑，强调一体化设计，是博地澜屋一直以来坚持的设计理念。四大主导专业相互融合、紧密配合、及时沟通，不仅保证设计理念贯穿整个项目的始终，更便于准确、高效的达到项目预期的最佳效果，包括设计方与业主达成的一致想法。而传统的建筑设计过程，往往因为缺失各专业间的密切联系，无形中增加时间成本，甚至出现反工现象。同时，我们相信在单科专业发展到一定程度的时候，在专业交叉领域的创新机会可能更多。因此，我们认为专业融合是一大优势，也必然是未来建筑设计领域的一个发展方向。

《新楼盘》：与一般的商业、办公建筑相比，总部科技园在建筑设计上有何不同之处？设计时要注意哪些方面的内容？

曹一勇：所谓的总部科技园一般是由多栋办公楼等商务功能建筑组合而成的一个成体系的、相对独立的区域。与一般的商业、办公建筑相比，除了满足基本功能需求外，它更需要满足企业和所有企业人的某种精神需求。

一方面，总部科技园的整体气质是企业形象的代表，它能够直观反映整个企业的性质、文化底蕴、包括业主的精神层次与追求。另一方面，它更强调"家"的感觉，注重园区内人的心理需求和归属感。也就是下面提到的"硬环境"和"软环境"并重。

《新楼盘》：在总部科技园的规划中，应该怎样处理好建筑与周围环境之间的关系？怎样去协调总部科技园的"硬环境"与"软环境"？

曹一勇：所谓的"硬环境"就是指园区规划内基本功能的需求。除商务办公以外，总部科技园内一般还有许多其他功能需求。比如我们设计的"北京东亿国际传媒产业园"项目，规划中除办公楼之外，还设有酒店、餐饮、会所、休闲、精品店，甚至邮局和银行等服务设施都一应俱全，这就代表了这个产业园的"硬环境"。

"软环境"则主要指文化底蕴、精神诉求这个层面，相对"硬环境"，它更加抽象，更需要我们用"心"去解读。同样以"北京东亿国际传媒产业园"为例，因为项目坐落于长安街东沿线，周边文化产业氛围浓厚，我们希望强调园区的文化性，在满足现代功能需求的基础上去创造具有文化底蕴、低调、内敛、经得起时间考验的建筑群。

整体规划中我们参考了北京传统四合院空间布局，提取本土建筑空间元素，强调"院"的概念。从初期规划设计就预留了空间，考虑到人、建筑、环境的交流与对话，同时为后期景观设计留下余地。建筑风格实际上是现代的，只是在色彩、材质的选择上我们提取了本土文化元素，包括景观元素的设计，思路都是一致的。

对总部科技园来说，设计过程需要处理好两方面的关系，缺一不可。我相信，一段时间之后，这类项目应该更耐看，人们不会以其新旧来评判它的价值，而是在精神文化层面上去接受和认可它，甚至这样的建筑本身就可以体现企业的文化品位和价值。

《新楼盘》：当下在总部科技园的规划建造中出现了一些盲目攀比的现象，比如过分的追求建筑外表的绚丽、建筑的体量规模等，您是怎么看待这个问题的？

曹一勇：我认为这是我国整个建筑领域普遍存在的一个问题，是盲目的求新、求异、求大的表现。建筑回归其本质就是一个遮风挡雨的庇护所，随着时代的发展，人们希望它在满足基本功能需求的同时还能带来精神层面的满足。我认为建筑仅仅是一个平台和载体，它最终服务的对象和主体是人，因此一个建筑不应该是特别的武断、特别的强势、特别不顾周边环境和人的感受而自成主体。

好的建筑标准可能各不相同，但是最终的目标是给人能够带来愉悦，而这种愉悦可以通过各种方式实现，比如说一般的乡村音乐和摇滚音乐，无论是通过抒情形式还是击打重金属的摇滚方式都能为音乐爱好者带来视听方面的享受。作为设计师、作为规划者和城市的管理者，不应该去一味的求新、求大，而忽略了人在空间的感受，不能局限于依赖视觉冲击去强调建筑本身。做一个建筑并不一定要去标榜什么，而是要实现其特定的价值。它，可以不奢华，但一定要恒久。

《新楼盘》：总部科技园在建筑设计的时候如何去实现其功能与形态的统一？

曹一勇：我认为建筑由低到高包含三个价值层面。首先是建筑的经济价值。建筑的投资者不管是个人还是政府，它一定要物有所值。取得一定的经济回报是不用回避的事实。比如一个住宅项目，如果设计得不好的话，不被社会和市场所认可，那么这个建筑一定是失败的。

在此基础之上，建筑具有社会价值。这方面更强调建筑的责任感，这也是建筑和其他艺术品最大的差别。艺术作品即使一时不被认可，也不会对社会产生消极影响。而建筑作品一旦建成，将不可回避的矗立在城市当中，直接影响到人们的日常活动，对城市形象、以及使用者的安全、便捷都有直接影响。它的社会价值还表现在它对环境的尊重、环保和可持续性。不计代价的建筑无论如何也不能称之为有责任感的建筑。所以建筑的社会价值与建筑师的社会责任心也是密不可分的。

最后即是更高层面的文化价值。建筑的发展历史与人类经济、文化发展史一定能够找到对应点。换句话说建筑的建造水平和形象一定反映了一个时代的经济水平和文化特点。建筑的文化价值是它对地域文化的表现与传承。建筑从精神层面给我们带来的感受与认知可以说是它的灵魂所在。

我希望设计的作品能朝着这三个方向迈进，合理的把握这几个方面，才能更好的实现建筑功能与形态的统一。

《新楼盘》：请简单谈一下您对当前总部科技园发展的看法，预测一下总部科技园未来的一种走向。

曹一勇：当今很多建筑都表现了一种无质量的增长趋势，可以概括为无"未来"的增长、无"情"的增长和无"根"的增长。

所谓无"未来"的增长，即不环保、不生态的增长方式。对于建筑来说就是要有绿色性和环保性，这也是未来科技园发展的方向之一。很多建筑在宣传节能环保与设计上是本末倒置的，比如建筑采用玻璃幕墙，然后再用各种新的技术去弥补这种不环保、不生态、不节能，这本身就是不对的。我一向主张建筑的"被动式"节能环保，不只是靠简单的技术堆积。简单的说就是环保≠高科技，有很多不必要的资源浪费在设计过程本身就可以解决。目前，生态节能和绿色建筑的观念已经被社会所认可，但是要真正务实理性的去实践这些内容，还有很长的路要走。

第二所谓无"情"的增长就是指经济在发展，但是人与人、人与周围环境之间的关系却愈发的淡漠。随着经济的发展，建筑的发展方向之一应该是通过建筑的设计、建造，为人们提供一个情感交流的平台。对总部科技园来说，就是要为工作生活在这其中的人们提供一个空间，这种空间与室内是明显不一样的，人在其中是一种放松的状态，人与人、人与环境之间是一种互动的关系。

第三就是无"根"的增长。所谓无"根"增长就是指随着经济发展，全球成了所谓的"地球村"，人们努力"求同"却淡忘了"存异"，整个过程使人们遗忘了历史，抛弃了地域文化和民族文化。如今，经济水平达到一定程度的时候，人们的精神需求被提上日程，建筑的文化传承问题已被重新解读。总部科技园今后的发展方向一定是要尊重地域文化，尊重传统文化。综合起来讲就是要有"未来"、有"情"、有"根"，这是未来建筑发展的一个趋势和方向。

如今住宅地产已经受到调控，曾经快进快出的卖方市场已不复存在。商业地产则表现相当的火爆，其竞争也日趋激烈。从广义层面讲，总部科技园也属于商业地产的范畴，其发展应该与今后地产投资商的开发操作思路有着密切的联系。随着商业地产进入"后商业地产"时代，其属性和产业地产越来越接近。所以说建设总部基地，期望"快进快出"、"短期利润翻倍"的这种思路一定是不现实的。未来总部科技园的发展，需要开发商、投资者或者后期的运营者抱着一种理性、客观、脚踏实地的想法去做开发、去培育和壮大市场，这样总部科技园才会得到一个良性、持续的发展。

BUILDING IS A PLATFORM FOR EMOTIONAL COMMUNICATION

—— Interview with Cao Yiyong, Chief architect of Beijing Building Life Architectural Design & Urban Planning Co., Ltd.

Profile:

Cao Yiyong
Senior Fellow of World Association of Chinese Architects
Member of APCREA Academic Committee
Expert Committee of CCREC
Chief Architect of Beijing Building Life Architectural Design & Urban Planning Co., Ltd.
National Registered Architect; Senior Engineer; Master of Architecture

Work Experience:

1990.7—2005.7 Chief architect of Design institute of Ministry of Posts and Telecommunications (later renamed: China Information Technology Designing Consulting Institute Co., Ltd.
2005.8—2010.1 Director and General Manager of Beijing Huate Architectural Design & Consulting Co., Ltd under 5+1 Werkhart International Group Co.)
2010.1— preset Chief architect of Beijing Building Life Co., Ltd.

Major Works:

Shantou Yuedong Information Building
Sichuan Mobile Headquarters Office Building
China Telecom Chongqing second Long Distance Telecommunication Building
China Telecom Lhasa second Long Distance Telecommunication Building
China Telecom Guiyang second Long Distance Telecommunication Building
China Telecom Chengdu second Long Distance Telecommunication Building
China Telecom Shanghai Information Industrial Park
Xinhua Bookstore Nanjing Logistic centre
Beijing Dongyi International Media Park
Chengde Museum
Henan University of Technology new campus
Yantai Zhifu Island Overall Planning
Beijing Spring mall Planning
Chengde Yudao Commercial street
Wendeng Nanhai Parfum—see
Chengde Shuangluan District People's Hospital
Tianjin Huanxiu Lake Courtyard Villa etc.

New House: As we know, infiltration and integration among architectural design, planning and design, landscape design, interior design is your unique advantage. Could you tell us what the important significance of this multiple targets integration is?

Cao: Taking into full account architectural design, planning and design, landscape design, interior design and stressing integrated design is the design concept that Building Life always insists. Above four leading disciplines integrating mutually, working closely, and communicating timely not only guarantee the design concept going through the whole project, but also high efficiently and precisely achieve optimum results that the project expected, including reaching an agreement between designer and owner. Nevertheless, in the process of traditional architectural design, due to lack of close connection between each discipline, it would cause time cost increasing invisibly or lead to rework phenomenon. In the meanwhile, we believe that when a single specialty develops to a certain degree, there will be more innovation opportunities in the cross—disciplinary field. Therefore, we consider that specialty combination is an outstanding strength which is sure to be a development direction in the future architectural design field.

New House: Compared with common commercial building and office architecture, what makes the Sci—tech Park different in the aspect of architectural design? And what should be specially paid attention to in the course of design?

Cao: Generally the so—call Sci—tech Park is a systematic and comparatively independent district composed by constructions and equipped with commercial functions like office buildings. In comparison with general commercial buildings and office architecture, apart from meeting basic function requirements, the sci—tech park could also be able to fulfill a certain spiritual need for business and business owners.
On one hand, the comprehensive quality of Sci—tech Park represents the corporate image, since it can reflect the entire enterprise' nature and background, containing owner's spiritual level and pursuit. On the other hand, it stresses more on 'home' feeling and focuses on people's psychological need and sense of belonging, i.e. below 'hard environment' and 'soft environment' are equal value.

New House: In Sci—tech park planning, how to deal with the relationship between buildings and surroundings? How to coordinate 'hard environment' and 'soft environment' in Sci—tech Park?

Cao: The so—called 'hard environment' refers to basic functional needs in site planning. Besides business office, generally there are also many other functional requirements. Taking the project 'Beijing Dongyi International Media International Park' we designed for example, apart from office building in this project planning, there are also other kinds of service facilities, hotel, catering, chamber, leisure, boutique, even post office and bank included. So this is the 'hard environment' in this industrial park.
'Soft environment' mainly means to cultural background and spiritual pursuit. It is more abstract compared to 'hard environment', and that is why it should be interpreted by heart. Again, for the project 'Beijing Dongyi International Media International Park', since it locates along Changan East Street around with rich cultural industry atmosphere, we would like to emphasize the cultural feature of this park, so as to create a architectural complex which possesses cultural background, low—key reservation, restraint and withstands the test of time based on meeting modern functional requirements.
In the overall planning, we refer to the spatial layout of Beijing traditional quadrangle then extract local architectural space element and underline the concept of 'courtyard'. Considering the communication among people, buildings and surroundings, we reserve space for it in the initial planning and design, meanwhile we also leave room for post landscape design.

As a matter of fact, such architectural style is modern, we just extract local cultural element from options of colors and materials, and moreover all the design and thought about landscape element is consistent with each other.

For Sci—tech Park, we have to coordinate the relationship between 'hard environment' and 'soft environment'. Because without one, the other suffers. I believe, after a while, this type of project would be more attractive. And people will accept and recognize it in the spiritual and cultural level rather than judge its value by whether it is new or old. What is more, such buildings could embody enterprises' culture status and value.

New House: Now some blind comparison phenomenon comes out in the course of Sci—tech Park project, for instance excessively pursuit of magnificent appearance and scale of buildings etc. So what do you think of it?

Cao: I think this is a common problem in China construction sector. It is a reflection of blind pursuit of the large, new and exotic. When it comes to the nature of buildings, it actually is a shelter for people to keep out the storm. With the development of times, people expect the building could bring satisfaction on the spiritual level while it meets basic function demands. Therefore, I reckon that building is just a platform and carrier, its final service object and subject is human. For this reason, buildings should not become a subject that is particular arbitrary, strong and ignoring the surroundings and people's feeling.

The standard of judgment whether it is a fine building may be different, but the final goal is to bring pleasure to people. And such pleasure could be achieved by various ways, such as country music and rock music, no matter it is through lyric form or through the rock and roll way of striking heavy metal, it could always bring enjoyment to music lovers. Similarly being a architect, a planner and city manager should not merely chase for new and large but ignore people's feeling in the space. Furthermore, to highlight architectures there are more ways other than simply relying on visual impact. To construct a building is not to flaunt something but to realize its specific value. So buildings should not be extravagant, but long lasting instead.

New House: When you are designing Sci—tech Park, how do you manage to integrate its function and its pattern?

Cao: In my point of view, buildings covers three value levels classified from low to high level. First of all, it is economic value that matters. No matter the building investor is individual person or government, the building must worth its value. And gaining a certain economic return is a fact that no needs to be avoid. For instance, if a residential project is poorly designed, it certainly could not be recognized or accepted by the public and market. Consequently this building fails.

On the basic of that, building also owns social value which more stresses on responsibility for building and this is the biggest difference between building and other artworks. Because artwork will not lead to negative impact on society even it is not recognized temporarily. However, once a building is finished, it is unavoidable to stand in the city and directly affects human's daily life, city image, users' safety and convenience. In addition, building's social value also shows in its respect to environment, environmental protection and sustainability. The building with no expensed—spared could not be treated as a responsible one in anyway. So buildings' social value and architects' social accountability is inseparable.

Last but not least is higher level, i.e. cultural value. It definitely can be found out corresponding points among building development history and human's economic and culture development history. In other words, construction level and image represents economic level and culture features of an era in a certain degree. Building's cultural value is the manifestation and inheritance to regional culture. In the aspect of spiritual level, feelings and recognitions brought by building is what the soul of building is.

I wish the designs advance towards mentioned above three directions and grasp above points reasonably. Only in this way can realize the integration between its function and its pattern.

New House: Could you briefly talk about your opinion of Sci—tech Park current development and make some predictions about the development trend of Sci—tech Park in the future?

Cao: At present lots of buildings show no quality growth trend that can be summarized as no 'future' growth, no 'emotion' growth and no 'root' growth.

So—called no 'future' growth means that the growth pattern is neither environmental—friendly nor eco—friendly. As for buildings, it should be with green quality and environmental benefits, and this is an important aspect in the future development trends of Sci—tech Park. However, many buildings are actually upside—down between designing and promoting energy efficiency and environmental protection. For instance, glass walls are applied to buildings then using different kinds of new technology to make up glass walls' defects like environmental—unfriendly, ecological system unfriendly, without energy efficiency. Designing in this way is completely wrong. I always stand for passive energy efficiency and environmental friendly for buildings which does not just simply rely on many technologies adoption to solve environment problems. In brief, environmental friendly is not equal to high technologies, and many sources are unnecessarily wasted in the course of designing while these sources could be saved by improving their designs. By now, energy efficiency and green building are accepted by society. In order to realize the practical, reasonable use of substances, there is still a long way to go.

Second so—called no 'emotion' growth refers that the relationship among persons, persons and surroundings is gradually changing to be apathetic, in the meanwhile the economy is progressing. Nevertheless with the development of economy, one development trend of buildings ought to provide platform for emotional communication to people. For Sci—tech Park, it is creating a room for people whose work life is in here. And this space is obviously different from inside door as people here are in a status of relax, so as to forming a kind of interactive relationship among persons, persons and surroundings.

Last is no 'root' growth. So—called no 'root' lies in the development of economy; the whole world turns out to be 'global village'. People try hard to 'seek commons' but forget to 'reserve diversities', they forget the history, abandon the regional culture and ethic culture over the process. Nowadays when economy develops to a certain degree, human's spiritual needs are on the agenda which lead to reinterpretation the culture inheritance issue of buildings. In future, the development trend of Sci—tech Park is sure to be respecting regional culture and traditional culture. All in all, combining 'future', 'emotion' and 'root' in architectures is definitely a development trend and direction for future buildings.

These days, residential property is already under control while the sellers' market which used to move fast does not exit any more. On the contrary, commercial property is in a great boom and its competition becomes increasingly fierce. Speaking in a broad sense, Sci—tech Park belongs to commercial real estate and its development should be closely linked to investors' ideas about future development operation. Along with commercial real estate stepping into 'post commercial property' period, its nature is getting close to industrial property. Therefore, it is unrealistic to build the headquarter base by counting on 'fast—moving' or 'benefits doubled in a short time'. The future development of Sci—tech Park requires developers, investors and future operators to explore, foster and expand the market by holding on to a rational, objective and realistic idea. In this way, Sci—tech Park would be developed healthily and sustainably at last.

EFFICIENT AND FLEXIBLE SPACE WITH THE CONCEPT OF SUSTAINABILITY

| ZOwonen Headquarter, Sittard
贯穿可持续理念的高效灵活空间 —— ZOwonen总部大楼

项目地点：荷兰斯塔德
业　主：阿姆斯特丹LSI项目投资公司
建筑设计：荷兰KCAP建筑与规划事务所
建筑面积：3 200 m²

Location: Sittard, The Netherlands
Client: LSI Project Investment N.V., Rotterdam
Architectural Design: KCAP Architects & Planners
Floor Area: 3,200 m²

项目概况

ZOwonen房产公司设在斯塔德的新总部建筑面积3 200 m²，将把公司的多个分公司集中于一座中心建筑中工作。本项目（总建筑面积24 000 m²）共包括5座建筑，将分阶段建成。而这座中心建筑是其中的第一座，符合GreenCalc 绿色建筑A级标准。

规划布局

新建筑是KCAP办公建筑群城市规划的一部分，属于阿姆斯特丹Rijnboutt b.v.总体规划的A部分。五座建筑环绕中心广场，随意散布于景观区。所有建筑的入口都朝向广场。

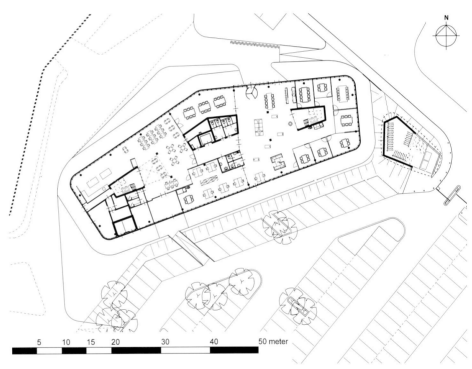

总平面图 Site Plan

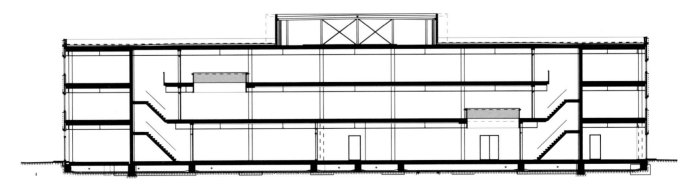

建筑设计

本设计的一个重要出发点就是日光可持续高效利用理念。立面设计按照建筑在城市布局中具体方位确定，41%的立面采用透明设计。立面采用横向全景窗，视野极其开阔，采光充足，反差小。立柱及稳定核心结构的设计，保证了办公空间使用的灵活性，从传统形式到开放的楼层布局皆可。建筑的最宽部分设有两个开口，保证阳光深入建筑内部，为不同建筑楼层间创造更多可见连接。

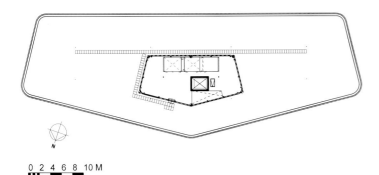
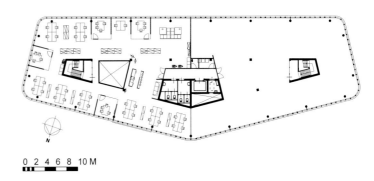
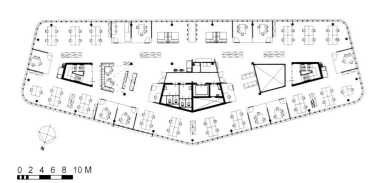

Profile

The new headquarters for ZOwonen housing corporation in Sittard comprise of 3,200 m^2 and will combine several of the corporation's branch offices in one central building. It will be the first of five buildings with a total of 24,000 m^2 to be realized in different phases. The building reaches a GreenCalc+ label A score.

Planning and Layout

The new building is part of KCAP's urban plan for the office cluster on part A within a larger masterplan designed by Rijnboutt b.v. from Amsterdam. The five buildings appear to be randomly distributed over the landscaped area surrounding a central square. All building entrances are oriented towards the square.

Architectural Design

An important starting point of the design was the concept of sustainability and the effective use of daylight. The design of the facades is dependent on its specific orientation within the urban setting, with 41% of the facade surface being transparent. The facade is designed with horizontal panoramic windows that not only offer magnificent views but also provide for higher daylight efficiency and less contrasts. The construction concept with columns and stabilizing cores allows for flexible use of the office space, from traditional to an open floor plan. Two voids in the widest part of the building let daylight penetrate deep into the building and create additional visual connections between the different office floors.

FEATURE | 专题

科技园建设要把握好规划的前瞻性与设计的灵活性
——访Zenx International Pty Ltd（哲思国际） 张奕和博士

■ 人物简介

张奕和
Zenx International Pty Ltd（哲思国际）设计总监/总建筑师
建筑设计博士，国内知名总体规划与建筑设计专家
哲思(广州)建筑设计咨询有限公司/深圳市哲思建筑设计咨询有限公司 董事长

1985年从华南理工大学硕士毕业后留校工作，1987年任华南理工大学讲师，1989年获澳大利亚国家奖学金到昆士兰大学攻读博士学位，1993年获博士学位。曾任职7年于澳大利亚最大的建筑设计事务所——peddle thorp architects, brisbane office（澳大利亚柏涛建筑设计事务所）的设计经理。2000年创建哲思国际，任总建筑师和设计总监。

张奕和博士有27年的规划与建筑设计的海内外项目经验，负责的大中型项目主要分布在澳大利亚和中国，涉及大中型商业零售、办公、居住及配套、科技与数码园区、文化教育、旅游及度假村开发等领域的总体规划与单体建筑设计。他对项目的独特理解、创新的设计构思为公司赢得了多个国际设计竞赛项目，已建成作品分布于中国的主要城市，并广受客户的好评。

《新楼盘》：哲思国际将时间、空间和人文看作是设计的最基本要素，那么设计师应该怎样在这个"三维坐标系"中定好位呢？

张奕和：这是我们总结了多年经验和很多实例而整理出来的理论框架。我们发现，每个建筑师在这三个向量中是有自己的不同取向，故而也可以把这作为一个评价体系来评价每一个作品。例如大家熟悉的现代派第一代大师柯布西埃（Le Corbusier）在时间上是着眼现代而反历史，在空间上着眼国际性而反地域化，在人文上是求大同而反差异；而跟他同一时代的另一名大师莱特（Frank L. Wright）则是用近乎于相反的取向。其实，一位负责任的建筑师在保持自己信念和风格的大取向前提下，对每一个设计任务也应该在这三个坐标上有不同的定位；因为有些项目可能更要求你尊重环境条件（空间矢量）多些，而另一些项目可能更多的需要考虑文化因素和功能需求（人文矢量）。我认为好的建筑师或者一个好的作品应该（时间上）与时俱进、（空间上）因地制宜、（人文上）因人而异。所以我们公司的设计理念和目标就是追求创造无愧于时代、同时尊重环境和文化的作品。

《新楼盘》：你们是如何理解总部科技园"总部"与"科技"这一概念的？

张奕和：各种类型的"科技园"在中国近年来如雨后春笋般涌现。其原型应该是国外的"Business Park"（商务园）和"Industry Park"（工业园），可以说是为科技办公和科技生产活动服务的集中园区。应该说，"总部科技园"是科技园中的最高端产品，它的出现来源于"总部经济"所催生的"总部基地"概念。它的定义有两方面特征：首先必须有跨国和跨地区的大型企业总部入驻在该区域形成集群布局；其次，这些企业的生产加工和配套基地通过其它形式安排在成本较低的周边和外地。所以我们所说的总部科技园可以说就是给大型科技企业量身定做的总部基地，承担的是这些大型企业的办公、研发、交流功能，建筑形

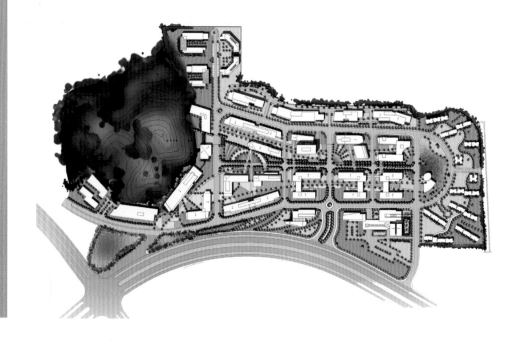

式是以独立的建筑物出现且企业有自己的命名权，是企业的形象和名片。总部科技园的面积可以小到几百平方米，大到几万平方米不等。

《新楼盘》："番禺节能科技园区"是你们设计的一个大型的总部科技园区项目，请简单介绍一下这个项目。

张奕和：ZENX进入中国的十多年来，已规划设计了十几个大中型科技产业园区。番禺节能科技园是我们早期规划、目前发展较成熟的一个大型园区，也是目前国内同一类型中较具影响力的一个。整个园区规划中有总部发展中心、创业中心、研发中心、培训中心、服务中心、科技产业带、白领公寓等功能组团，项目占地50公顷，建筑面积70万平方米，以前瞻性的规划，先进的设计，立体的生态空间和智能化的信息网络实现了生态与智能完美结合，科技与人文交相辉映。缔造了科技园区的国际典范，开启了财智时代生态办公的新境界。去年胡锦涛总书记考察广东时，还专门视察了该园区，良好的园区规划环境、符合当地人文地理的建筑设计在国内引起了很大反响。从2004年参与园区总体规划开始，我司已完成了1—6期的总体规划与建筑设计，目前我司正在规划设计园区第7、8期的开发，基本上是本园区的收官之作。

《新楼盘》：作为一个节能型的科技园区，其在选址和环境方面是否有特定的要求？在前期规划的时候需要考虑哪些因素？

张奕和：一个大型的科技园区，在选址上是一个多因素多方面的综合决策过程，不能单从建筑和规划上考量。这一方面，政府、开发商、规划师一定要紧密配合。由于对大量土地面积的需求，这一类园区通常需要在郊区或城乡结合部，所以它的地理位置、交通条件在很大程度上影响了后面的招商和入园企业以后的日常运作。自然环境条件同样也很重要：对很多大型企业来说，选择在产业园里办公，比在城市中心办公，其中主要的优点之一就是环境优美，当然还有停车方便、租金低廉等等其它因素。所以要尽量利用自然环境条件。番禺节能科技园早期规划的时候就定下了保留原来基地内的小山包，利用它做大文章的理念，不但前期的建筑物围绕它展开，更采取引"绿"下山形成中轴最

平的手法使绿色从山上下来流到每一个组团中（插图）。几年下来，目前已能看到这一先见的成果和效应了。此外，基础设施对于园区吸引企业也很重要，企业的交通、住宿、生活配套等园区往往也是民企的大难题，一定要规划在先、独步完善。番禺科技园在第三期的时候就开始配备小型会所，四期的时候配备了员工公寓。目前的8期更增设小型交流中心等等，银行和餐饮则随着发展由市场机制自然形成。

《新楼盘》：与普通的科技园区相比，番禺节能科技园区在规划与建筑设计上有什么特点和不同之处？

张奕和：番禺节能科技园规划上的一大特点是生态、低碳。基地刚开始可以说是一片空地、一张白纸。但通过保留原有山包、引绿入园、人车分流空出绿地，几年下来后，目前已是绿色满园、花香鸟语。当然，平均1.5的容积率也帮了不少忙。另外一个特色就是规划上预留的灵活性。一个大型园区，建设周期很长，对市场的变化要有前瞻性，从规划上也要预留可能的变化空间。值得一提的是，番禺科技园的甲方在规划初期只是把她定义为综合的科技园区，并未考虑总部楼产品，只是随着园区的发展和成功，很多大型企业和上市公司陆续进驻，市场才提出了这一块的需求。因此，在刚完成的7—8期规划中，总部楼占了很大的比例和份量。建筑设计上，我们也给予多种需求最大的灵活性。产品涵盖了入园起步企业的"孵化"产品到"总部基地"产品等多种类型。面积区间小到一百多平方米的分户到几千平方米的独栋总部楼，建筑形式也因此而丰富多样。此外，在建筑技术上，也采用一系列措施支持企业的运作需求，例如小单元用分体式空调系统，总部楼和大单元采用VRV系统，让企业可以全天24小时随时灵活使用。建筑处理上则在统一的现代亚热带风格下灵活多变，形成了整个园区目前既统一又有变化的风貌。

《新楼盘》：当下建筑设计领域的绿色节能理念不断被重视，请谈一下总部科技园在这方面的创新以及发展前景是怎样的？

张奕和：随着中国经济的发展和中国在全球经济和政治地位的提升，总部园这块的产品肯定会继续发展壮大。但产品自身也需要更新换代升级，应时应需而变。节能主题会是重要创新方向之一。但这一块需要政府政策的鼓励扶持、市场/社会节能意识的提高、发展以及设计师的共同努力。作为规划设计师，我们有责任推进这一领域的工作，但要循序渐进，选择市场成熟度高、相对可实施的技术先行，一步一个脚印的走。我相信，最终应该是会达到我们的目标的。

The Foresight of Planning and Flexibility of design should be held for Sci—tech Park Projects

—— interview of Doctor Zhang Yihe of Zenx International

Profile:

Zhang Yihe
Design Director and Chief Architect in Zenx International Pty Ltd
Doctor Degree in Architectural Design
Famous Master—plan and Architectural Design Specialist in China
Chief Executive of Zenx Guangzhou/Shenzhen Architectural Design and Consultant Co., Ltd

Zhang Yihe acquired his Masters degree in 1985 from South China University of Technology (SCUT) and started to work as a teacher there. He taught in SCUT as an instructor in 1987. He won an Australian National Scholarship in 1989 and went to study his PhD at the University of Queensland, which he completed in 1993. Zhang worked at the highly—regarded Australian architectural practice Peddle Thorp Architects in Brisbane for seven years before he founded Zenx International in 2000. He is Chief architect and design director in Zenx International.

With 27 years of experience in planning and architectural design, Zhang Yihe's projects are mainly in Australia and China. They cover master planning and architectural design of large and medium—sized commercial retail, office buildings, residential and supporting facilities, sci—tech parks and digital parks, cultural and educational buildings, tourist projects and vocational village development, etc. With his distinctive understanding and creative design concepts, he has won several awards through international design competitions. Zhang's finished projects are located throughout major cities in China and he is held in favorable reputation by his clients.

New House: In Zenx International, time, space and humanity are regarded as the basic factors of design, so how should architects set their orientation in this three—dimensional system?

Zhang Yihe: This conceptual framework has undergone the test of years of experience and practice. As for what we have discovered, every architect has his own orientation in these three fields, which could also be taken as an assessment system for their work. Take Le—Corbusier, master of the first generation of the modernist school, for example. He followed modernity, opposed history in time, pursued internationalism, opposed regionalism in space, welcomed great unity and opposed differences in humanity. However another master of his time, Frank Lloyd Wright, adopted almost the opposite view to Le—Corbusier. In fact, responsible architects can have different orientations for each single work in this system as, while sticking to their own belief and style, some projects may require more respect to environmental conditions or space vectors while other projects may need consideration of cultural elements, functional elements or humane factors. Therefore in my point of view, I think a good architect or good design work should advance with the times, suit the local circumstances and vary for different people. That is why we want our projects to live up to time, respect the environment and value cultures, which are the overall design concepts and goals of our company.

New House: How do you understand the words Headquarter and Sci—tech in Headquareter Sci—tech Park?

Zhang Yihe: Different types of sci—tech parks have sprung up like mushrooms in China in recent years. Headquarter Sci—tech Parks refer to a centralized area which serves scientific and technological work and production activities, with prototypes being Business Parks and Industrial Parks in the west. Headquarter Sci—tech Parks should be viewed as the highest—end product of its kind, whose appearance originates from the concept of a headquarters base within a corporate economy. The definition implies two features: firstly, there should be large international or inter—regional enterprises which form cluster pattern in the area; secondly, these enterprises have their manufacture and supporting facilities set up in low cost surrounding areas or other places. Therefore Headquarter Sci—tech Parks are the headquarter base tailored for large sci—tech enterprises, which afford functions like office work, R&D and exchanges. As the image and business identity for these enterprises, Headquarter Sci—tech Parks take the form of independent buildings, for which enterprises have their own naming rights. The area of Headquarter Sci—tech Parks could be as small as hundreds of square meters or as large as tens of thousands of square meters.

New House: Panyu Energy Conservation Sci—tech Park is one of the large Headquarter Sci—tech Parks you designed. Please introduce this project a little bit.

Zhang Yihe: In more than ten years since Zenx entered the Chinese market, we have designed scores of large and medium—sized sci—tech parks. Panyu

Energy Conservation Sci—tech Park, which we planned early, is a large—sized park with mature development and is an influential development of its kind in the whole country. Covering an area of 50 hectares, with floor area of 700,000 square meters, the sci—tech park has a planned headquarter development center, entrepreneurship center, R&D Center, training center, service center, sci—tech industrial area, white collar apartments, etc. With its foresight in planning, advanced design, ecological space and intelligent information network, the project has achieved the perfect combination of ecology and intelligence, technology and humanity, which has also forged a good international example of sci—tech parks. It has opened up a new realm for ecological work within this era of finance and information exchange. Last year, when President Hu Jintao came to inspect Guangdong, he also went to Panyu Energy Conservation Sci—tech Park. The beautiful planning environment and architectural design which enhance the local cultural and geographic context caused a great sensation in the whole country. Since the overall planning started in 2004, our company has finished the general planning and architectural design of Phases 1 to 6. At present we are working on the planning of Phase 7 and 8, which are basically the final stages of the project.

New House: As an energy conservation sci— tech park, are there any specific requirements for site selection and environment? What factors have to be taken into account in pre—planning?
Zhang Yihe: It is a comprehensive process involving various factors and aspects in the site selection for a large—scale sci—tech park. It should not be decided simply by architecture and planning, as it requires the close coordination among the government, developers and architects. This type of park will be located in outskirts and marginal urban areas because of their requirement for large—sized land area, so the location and traffic conditions play a great role in investment attraction, future operation and running of the enterprises. The natural environment conditions are also important; most large—scale enterprises set their offices in the industrial park rather than in city center for its graceful environment mainly, as well as the convenient parking and lower rent cost. For this reason, natural environment condition should be put into best use. In the pre—planning of Panyu Energy Conservation Sci—tech Park, a concept was that the existing hill was to be retained and taken full advantage of; the resulting architecture has been set surrounding the hill and the greening will be extended downhill to every complex. Over the past few years, the achievement and effect of this concept has obviously appeared. Furthermore, the facilities and infrastructures play an important role in attracting enterprises moving in. The tough problems for private enterprises such as transportation, accommodation and living support should be completed in the pre—planning. The Third Phase of Panyu Energy Conservation Sci—tech Park has been equipped with small—scale clubs while the Fourth contains staff apartments. The Eighth at present has been installed with the small—scale communication centers etc, banks and F&D emerging along with the free market system.

New House: Compared with the normal sci—tech parks, what's the characteristics and difference in the planning and architectural design of Panyu Energy Conservation Sci—tech Park?
Zhang Yihe: One main point in the planning of Panyu Energy Conservation Sci—tech Park is its ecology and low—carbon concept. The base is like an open space, a blank sheet in the very beginning, but it has turned out to be a park with greening, flowers and birds after several years by retaining the hill, extending the greening and setting the landscape space between pedestrians and cars. The plot ratio of 1.5 on average counts for a lot. Another characteristic is the flexibility in the planning. In the planning of such a large—scale park with a long—construction period, architects should be forward—looking about market changes and reserve space for potential changes. It is worth mentioning that the owner of the project defined it in the pre—planning as a comprehensive sci—tech park without the headquarters buildings; with the development and success of the park, with a lot of large—scale enterprises and listed companies moving in, the market consequently called for that. Headquarters buildings have occupied a high ratio in the recently completed Seventh and Eighth Phases. We maximised flexibility to various demands in the architectural design, with our planning to suit various types of buildings like the incubation buildings for the take—off enterprises and the larger headquarters bases, etc. The architectural layouts vary from apartments of more than $100m^2$ area to detached headquarters building of thousands of square meters. Besides, a series of measures supporting the enterprises' operations have been adopted in the architectural techniques, such as the separated air—conditioning system for small units, RVB System for headquarters buildings and large units, to guarantee enterprises 24—hours flexibly and smooth running. The whole park has presented variation within unity by adopting various modern subtropical styles.

New House: Given that the concept of green and energy saving has been paid constant attention in the architectural design field, could you talk about the innovation and development prospect of headquarters sci—tech parks on this aspect?
Zhang Yihe: With the development of the Chinese economy and the promotion of China's status in global politics and economy, the headquarters park projects will continue their further development. However as a product it requires updating and regeneration according to the changes of trends and demands. The energy conservation issue will be one of the necessary directions in innovation which could not move on without the support of government policies, the enhancement and development of energy conservation awareness in the society and market, and the joint effort of architects. As planning architects, we are responsible to carry forward projects in the field step by step, beginning with the executable techniques in the market framework and building a collective body of knowledge that I believe will lead to our goals eventually.

INTEGRATION OF SUSTAINABILITY, CREATIVE DESIGN & INNOVATIVE TECHNOLOGY

| Panyu Tian An Hi-tech Ecological Park

生态、创意与新科技相融合的生态型科技园 —— 番禺节能科技园

项目地点：中国广州市番禺区
开 发 商：广州市番禺节能科技园发展有限公司
规划设计：Zenx International Pty Ltd（哲思国际）&
　　　　　DEM Design
建筑设计：Zenx International Pty Ltd（哲思国际）
占地面积：509 300 m²
建筑面积：713 644 m²

Location: Panyu, Guangzhou, China
Developer: Panyu Energy Conservation Science&
Technology Park CO. Ltd
Planning and Layout: Zenx International Pty Ltd &
DEM Design
Architectural Design: Zenx International Pty Ltd
Land Area: 509,300 m²
Floor Area: 713,644 m²

项目概况

本项目位于广州市番禺区西北侧，迎宾路与北二环路交汇处，紧邻番禺未来的核心地段——番禺新城CBD，且周围环境优美，拥有极佳的自然景观资源。

作为迎宾路上的标志性建筑群，整个园区在与周围环境紧密结合，强调节能环保理念的同时，也彰显个性，充分展示园区建筑立面变化统一，色彩丰富多样的特点。本着技术领先、规划率先、环境优先的原则，本项目现已发展成为国内可持续发展科技园的典范。

规划布局

本项目占地509 300 m²，拟建建筑面积713 664 m²，规划中延伸西面山体使之融入中央主轴线，使之形成整个科技园的绿色主旋律，沿主轴线布置三个主要空间节点：一个中心中央广场，两个副中心；绿色论坛和滨湖广场。沿主轴线两侧，营造舒适宜人的连贯院落式室外休闲带，改善工作环境和调节微气候。中央林荫道北侧布置大尺度的"千步廊"作重点绿化/遮阳元素，起连通西面山体与湖滨广场的纽带作用。规划中通过对项目场地的解读，组织高效能低干扰的（人行/车行）交通系统。

FEATURE | 专题

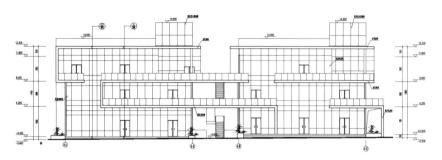

立面图 1 Elevation 1

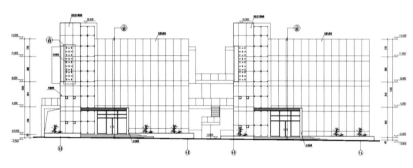

立面图 2 Elevation 2

建筑设计

园区的建筑风格采用与广州气候特点相似的澳大利亚亚热带建筑设计风格，在建筑色彩上，大胆运用红、黄等跳跃的色彩，配合茂盛的景观植被，令整个园区充满色彩缤纷的阳光气息。园区的产业楼建筑类型主要分为高层的产业大厦和低层的总部楼两种。产业大厦考虑到地下停车效率及研发单元的实用性，一般都采用8.4 m柱跨，单元面积在150~500 m²之间。为体现生态节能的特点，所有电梯厅都能自然采光通风，并配备有观光电梯。考虑到尽量避免货流和人流之间的互相干扰，一般在设计中考虑将办公流线和后勤货运流线各自独立。公共交通面积较少，实用率高达85%，并且每套单元都带有独立卫生间。为体现亚热带建筑特点，多数单元都配有阳台，为单元提供室外休闲和布置绿色植物的空间，充分体现建筑的人性化设计。

总部楼一般都由大企业单独使用，为保证单元空间的高实用性，采用大跨度柱网。建筑配备有独立的庭院，顶层有休闲露台，部分高档总部楼还拥有中庭空间，对楼内采光和调节室内微气候起到很好的作用，建筑空间也因此变得生动有趣。东西外墙材料采用干挂陶板和铝板，南北面采用双层LOW—E玻璃及设置阳台，在考虑美观的同时，关注其功能的实用性。立面以虚实对比的手法，利用阳台及屋盖的立面线条，营造出生动活泼的视觉效果。

Profile

The project is located in the northwest of Panyu, at the intersection of Yingbin Road and North Second Ring Road, and closely adjoins the future new town CBD of Panyu with superior surroundings and rich natural landscape resources.

As a landmark building group, the project has a close relationship with the surroundings and puts emphasis on energy conservation and environmental protection. The project keeps its own personality to express a balanced composition of color variations and facade elements. Using the principles of technological advancement, forward-planning and environmental priority, the project has developed into a model civil high-tech park with sustainable development.

Planning and Layout

The project has a land area of 509,300 m^2 and a proposed floor area of 713,664 m^2, infusing the

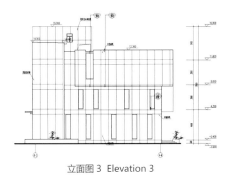

立面图 3 Elevation 3

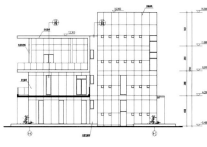

立面图 4 Elevation 4

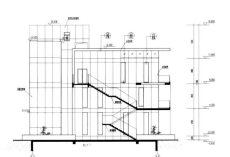

1—1 剖面图 1—1 Section

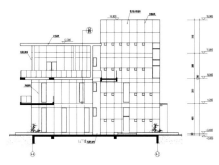

2—2 剖面图 2—2 Section

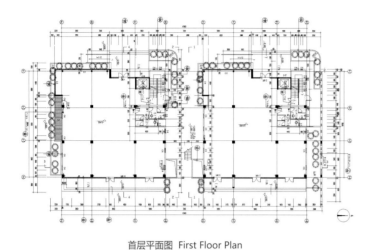

首层平面图 First Floor Plan

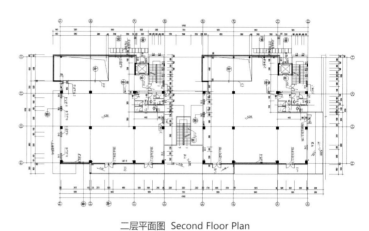

二层平面图 Second Floor Plan

mountain in the west into the central axis line to shape the theme of the whole high-tech park. Along the central axis line, there are three space nodes. One is the central square and the other two are sub-centers: Green Forum and Binhu Plaza. On both sides of the central axis line, the comfortable courtyard leisure belt improves the working environment and adjusts the microclimate. Thousand Step Corridor in the north of the central avenue services as a bridge to connect the mountain in the west and Hubin Plaza. Such planning and layout helps to organize a highly-efficient and low-interference traffic system.

Architectural Design

The design adopts the Australian subtropical architecture style, boldly using red and yellow as the main colors inspired by lush tropical vegetation and sunshine. The industry buildings mainly include high-rise offices and low-rise headquarters buildings. High-rise industry offices focus on the practical use of the underground parking and R&D units, so such buildings usually adopt an 8.4m column grid and unit areas ranging from $150m^2$ to $500m^2$ In order to show the qualities of ecology and energy conservation, all the elevator halls can receive natural lighting and ventilation and are equipped with sightseeing lifts. The offices streamline the logistics of goods loading, which is separated individually to avoid interference between goods and people.

The public areas are relatively small with an efficiency ratio of 85% and every unit has an independent bathroom. To reflect the subtropical architecture characteristics, most units have a balcony for outdoor leisure and planting.

Headquarters buildings are owned exclusively by large enterprises. They apply long-span column grid to guarantee the practical use of space. There are independent courtyards, leisure terraces at the top floor and even atriums in some top-grade headquarters buildings for interior lighting and microclimatic benefits. The east and west external walls adopt ceramic and aluminum plates while the south and north facades use double-glazed Low-E glass and shaded balconies. The facades take advantage of balcony lines and roof forms to create a vivid and lively visual effect.

FEATURE | 专题

LANDMARK BUILDING INSPIRED BY FREEDOM UNDER THE SLAT COVER

| Head offices of the Telecommunications Market Commission, CMT
板式表皮覆盖下自由催生的标志性建筑 —— CMT总部大楼

项目地点：西班牙巴塞罗那	Location: Barcelona, Spain
客　　户：Groupo Castellvi	Client: Groupo Castellvi
建筑设计：Batlle i Roig Arquitectes	Architectural Design: Batlle i Roig Arquitectes
设 计 师：Enric Batlle, Joan Roig	Architects: Enric Batlle, Joan Roig
合作设计：Goretti Guillén, Meritxell Moyá, Helena Salvadó, architects, G3 Arquitectura, STATIC, Gerardo Rodríguez, PGI Grup SL	Collaborators: Goretti Guillén, Meritxell Moyá, Helena Salvadó, architects, G3 Arquitectura, STATIC, Gerardo Rodríguez, PGI Grup SL
面　　积：12 000 m²	Area: 12,000 m²

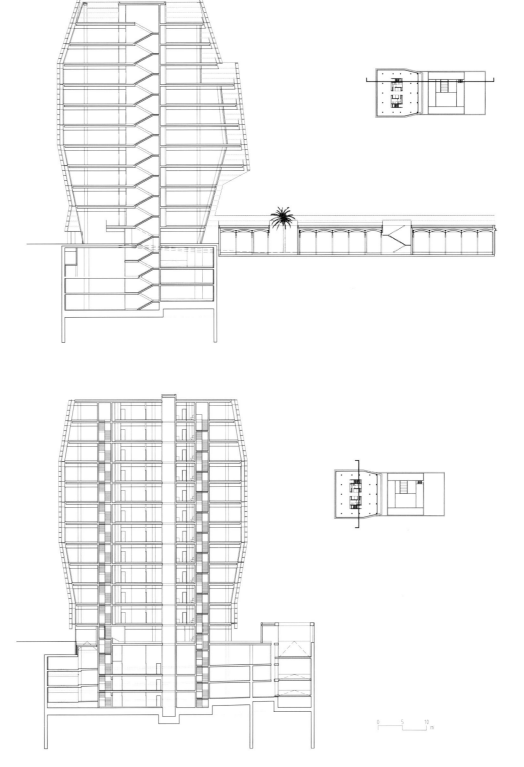

项目概况

CMT总部大楼位于22@区,和INTERFACE办公大楼一起,与周边的建筑共同组成一个建筑群,包括由Grupo Castellví开发的22@商业园和超过41 000m²的综合性办公及酒店建筑。它与四个街区相联:Bolivia、Ciutat de Granada、Sancho d'Àvila 和Badajoz。这一区域也是工业遗产保护区,有着1906年建立的Can Tiana纺织厂。

规划布局

CMT大楼在一个狭长的用地之中,主要立面面向Carrer Bolivia。老的工业厂坐落在场地中间,项目的出发点正是要将其融入CMT的功能规划中。建筑包含地下3层停车场以及11层的办公服务区域。底层作为建筑的进出通道,与原来的旧厂房相连。旧厂房在原有结构基础上被改造成330人的大会议室以及CMT员工服务礼堂,厂房屋顶也与大楼的地面楼层相连接。

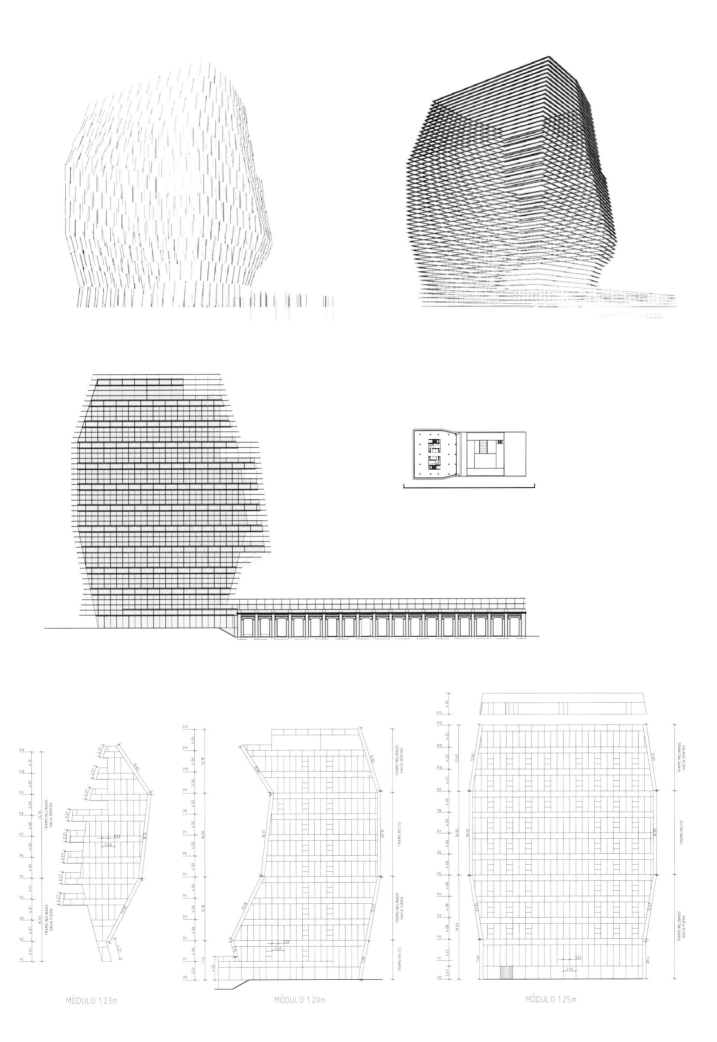

FEATURE | 专题

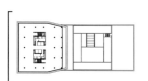
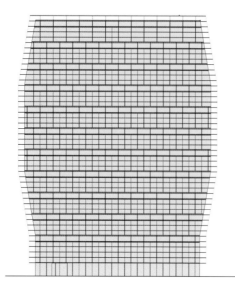

建筑设计

包括办公区域在内的主要建筑以入口和服务区为中央核心，办公区域在其外延四周，充分利用建筑内部的自由空间，向四面开放。这样的设计使其建筑整体呈现细长不对称状，区别于周边的其他建筑。这种自由的形式又催生了奇异的体量，离散了建筑的表面，不光影响着建筑的内部空间，也使得建筑成为一个标示性的建筑。

建筑师在新建筑上采用与老厂房一致的水平板式表皮系统，将两个建筑统一起来。板式表皮系统能为每层遮阳，同时也形成了地面入口的雨棚。

NEW HOUSE _077

Profile

Like the INTERFACE building, the building for the Telecommunications Market Commission (CMT) is situated in the 22@ district. In this case, the building forms part of a complex, 22@ Business Park, developed by Grupo Castellv í and containing a large business complex of over 41,000 m² of offices and hotels, delimited by four streets: Bolivia, Ciutat de Granada, Sancho d'Àvila and Badajoz. Also set in this sector is one of the factories of the old Can Tiana textile mill, built in 1906 to a project by Guiteras and listed in the Special Protection Plan for the Industrial Heritage of Poblenou.

Planning and Layout

The CMT building stands on a long, narrow site that presents its main facade to Carrer Bolivia. One of the old Can Tiana factory buildings stands at the centre of the site, and the project sets out to recover and incorporate it into the CMT's functional programme. The main volume comprises three basement floors for car parking and eleven floors above grade with offices and services. The ground floor, providing the function of access and entrance, connects with the old mill building, the original structure of which is conserved as an auditorium with capacity for 330 persons, a large meeting room and services for CMT employees. The roof of the mill was adapted for use and connects with the first floor.

Architectural Design

The main volume, containing the offices, is organized around a central nucleus of entrances and services, and the workstations are laid out around it, making full use of the spatial freedom enjoyed by a building that opens out on all four sides. This allowed us to explore a volumetric formalization that sets the building apart from its surroundings, highlighting its lengthwise asymmetry. This formal freedom produces a singular form, faceting the faces of the building and moulding it as a unique, recognisable piece that finds its reason for being in an innovative relation between exterior and interior.

The decision to bring a unitary treatment to the building's outer appearance led us to protect its facade using a horizontal slat system throughout its volume that continues over the old factory, connecting the two. The slats serve to cover the upper terraces and installations, and form an awning at the ground floor entrance.

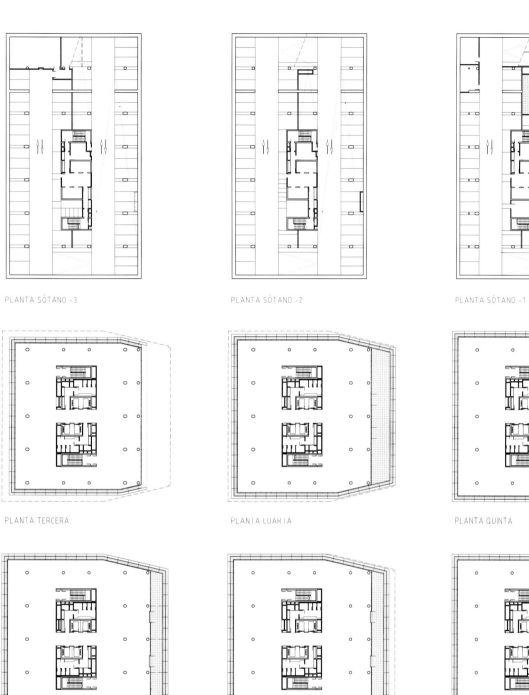

PLANTA SÓTANO -3   PLANTA SÓTANO -2   PLANTA SÓTANO -1

PLANTA TERCERA   PLANTA CUARTA   PLANTA QUINTA

PLANTA SEXTA   PLANTA SÉPTIMA   PLANTA OCTAVA

PLANTA NOVENA   PLANTA DÉCIMA   PLANTA TÉCNICA

CONCISE AND DYNAMIC POSTINDUSTRIAL PARK

| Hangzhou Bay Science & Technology Center

简约而充满动感的后工业化园区 —— 杭州湾科技创业中心

项目地点：中国浙江省慈溪市
业　　主：杭州湾开发区管委会
建筑设计：DC国际
总用地面积：28 950 m²
总建筑面积：39 942 m²
容积率：1.23

Location: Cixi, Zhejiang, China
Client: Administration Committee of Hangzhou Bay Development Zone
Architectural Design: DC Alliance
Gross Land Area: 28,950 m²
Gross Floor Area: 39,942 m²
Plot Ratio: 1.23

项目概况

基地位于慈溪市杭州湾新区内，紧邻杭州湾大桥。而慈溪这个城市有着悠久的文化历史和独特的地理位置。它位于东海之滨，东离宁波60 km，北距上海148 km，西至杭州138 km，是长江三角洲经济圈南翼，环杭州湾地区上海、杭州、宁波三大都市经济金三角的中心，区位和交通优势十分明显。

规划布局

同如今的大多数公建项目一样，业主对建筑的标志性做出了要求，希望其成为地区的中心，代表这一区域的城市形象。而且，其达到64 m的高度使其有可能对周边地区起到强有力的控制作用。这种标志性其关键的物质特征是具有单一性，或在某些方面具有唯一性，有清晰肯定的形式，与背景形成强烈的对比。在本项目中，这种单一性与唯一性是由大量的单元在自由聚集的过程中形成的，单元的叠加带有某种自发的形式趣味，最终将生活的秩序反映到建筑的城市形象中，通过大尺度的表达成为特别的城市标志。

建筑设计

在造型设计中，集中体现了现代建筑的美，以现代、简约、充满动感的风格形成了清新而鲜明的个性，给人以强烈的感染。在形体处理上注重高低错落与体型的变化，办公楼突出体现建筑的体量感及办公建筑的特性，材料以玻璃、面砖、质感涂料、压型钢板为主。利用有限的造型元素进行变化，使建筑立面形成流淌的音乐般的韵律美。

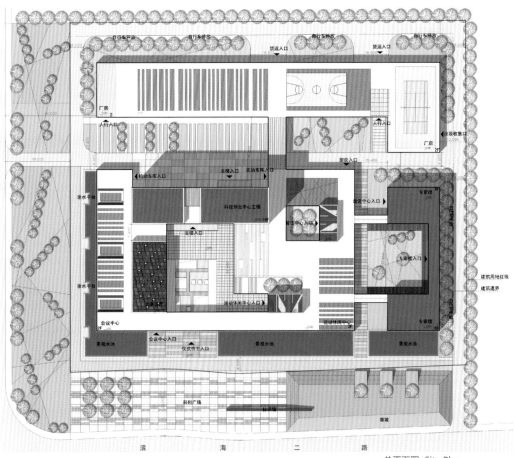

总平面图　Site Plan

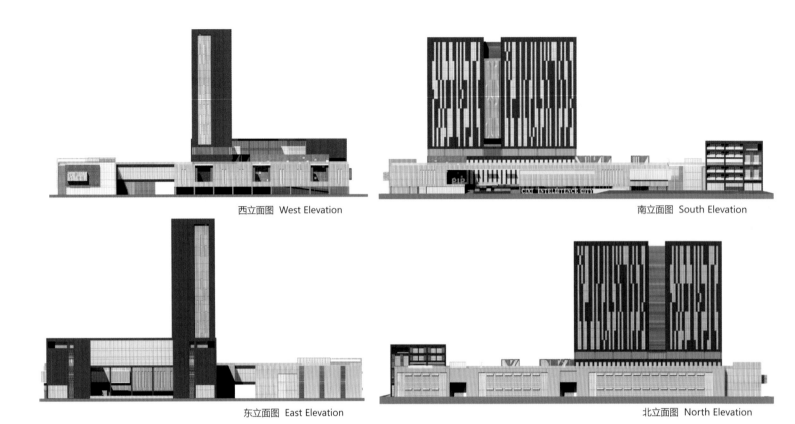

西立面图 West Elevation
南立面图 South Elevation
东立面图 East Elevation
北立面图 North Elevation

Profile

The project is located in Hangzhou Bay New District next to Hangzhou Bay Bridge in Cixi City with long—standing culture and history as well as peculiar geographical position. The city is in the shore of the East China Seam, 60 km away from Ningbo in the east, 148 km from Shanghai in the north, 138km from Hangzhou in the west. It is the south wing of the economic circle in the Yangtze River delta, adjacent to the center of economic and financial triangle composed by three metropolis Shanghai, Hangzhou and Ningbo in Hangzhou Bay Region, occupying a superior position and traffic advantage.

Planning and Layout

Like most recent public architecture project, the client has the requirement about the architecture that it will be iconic to be the center in the region, as well as the urban icon and image. Furthermore, the 64m high architecture is likely to bring the dominant effect on the surrounding regions. The

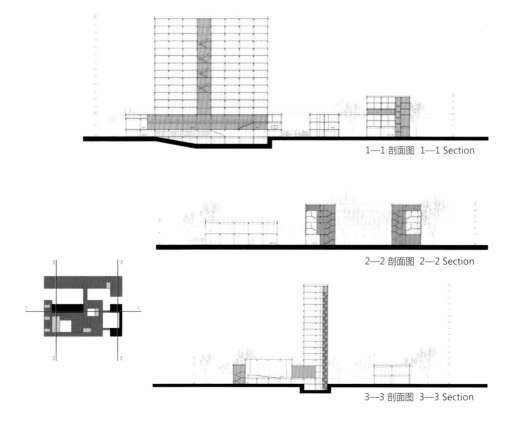

1—1 剖面图 1—1 Section

2—2 剖面图 2—2 Section

3—3 剖面图 3—3 Section

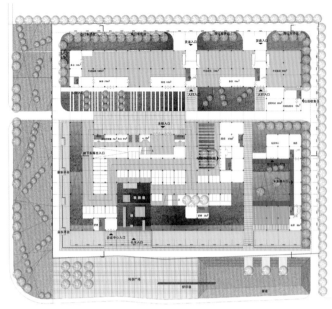
一层平面图 First Floor Plan

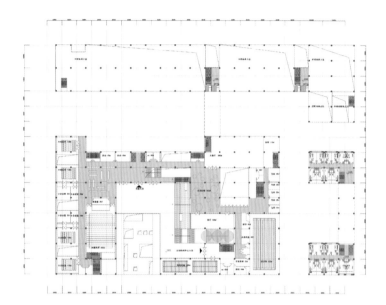
二层平面图 Second Floor Plan

三层平面图 Third Floor Plan

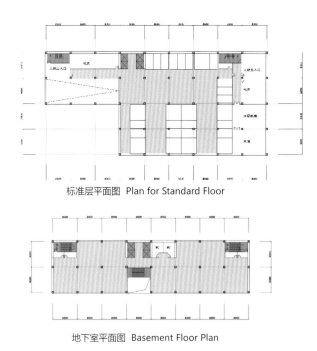

标准层平面图 Plan for Standard Floor

地下室平面图 Basement Floor Plan

key iconic characteristic is unity or uniqueness in some aspects, with strong comparison to the background in definite and clear form. In the project, the unity and uniqueness is generated in the free aggregation of a large quantity of units superimposed in interesting pattern. Finally the living order is reflected into the urban image of the architecture to be the special urban icon by powerful expression.

Architectural Design

The beauty of modern architecture is epitomized in the shape design, forming the fresh and vivid characteristic in modern, concise and dynamic style, presenting the captivating effect. The scattered levels and the shape changes require attention in dealing with the building shapes. The office building highlights the construction volume and the office architecture characteristic with major materials of glass, face bricks, texture coatings and profiled steel plates. The facades of the building present the fluent musical rhythm by the changes with limited shape and style elements.

BUILDINGS CREATE EMOTIONAL COMMUNICATION PLATFORMS AMONG PEOPLE | Beijing Dongyi International Media Park

建筑铸就人与人之间情感交流的平台——北京东亿国际传媒产业园

项目地点：中国北京市	Location: Beijing, China
业　　主：北京中视东升文化传媒有限公司	Owner: Beijing Zhong Shi Dong Sheng Culture & Media Co., Ltd
建筑设计：北京博地澜屋建筑规划设计有限公司	Architectural Design: Beijing Building Life Architectural Design & Urban Planning Co., Ltd
用地面积：56 926.67 m²	Land Area: 56,926.67 m²
总建筑面积：181 570.19 m²	Total Floor Area: 181,570.19 m²
地上建筑面积：155 727.13 m²	The Ground Floor Area: 155,727.13 m²
地下建筑面积：25 843.06 m²	The Underground Floor Area: 25,843.06 m²
容 积 率：2.7	Plot Ratio: 2.7
绿 化 率：47%	Greening Ratio: 47%

项目概况

北京东亿国际传媒产业园坐落于北京长安街东延长上，东临五环路，西瞰兴隆公园，北依朝阳路，南至京通快速路，处于北京CBD—定福庄传媒走廊的核心位置。园区建筑布局规整，以独栋多层为主，整体风格简洁大气，并赋予文化气息的传达，在景观方面注重打造自然式园林景观与家庭式办公环境，实现真正意义上的OFFICE PARK。

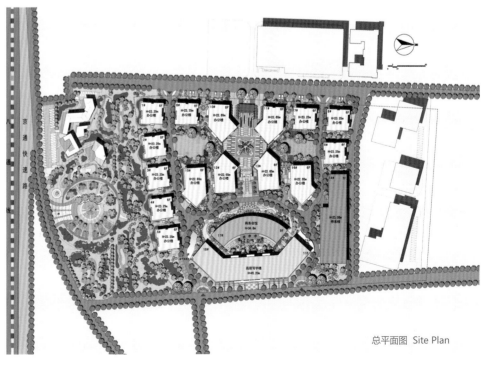

总平面图 Site Plan

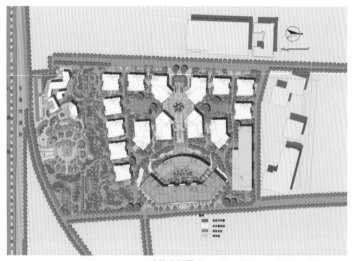

功能分析图 Functional Analysis Drawing

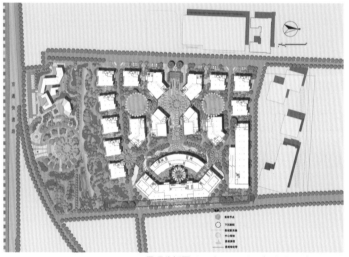

景观分析图 Landscape Analysis Drawing

规划布局

园区内"工"字道路把园区分为四个部分，北面是停车库和两个办公楼；西面中心布置了商业和办公楼的综合体；在东面中心设计的是整个园区之首的高层写字楼和酒店；南面则是五栋独栋办公楼。每个地块建筑独具特色，紧紧相依。东面的高层写字楼和酒店环抱相依，将整个建筑群凝聚到一起。

建筑设计

项目规划中参考了北京传统四合院空间布局，提取了本土建筑空间元素，强调"院"的概念。设计师从初期规划设计就预留了空间，考虑到人、建筑、环境的交流与对话，同时为后期景观设计留下余地。

建筑采用现代简约风格，运用中式经典色彩，深灰、浅灰、红与一期建筑色彩相呼应。内部开敞办公空间，顶层露台，外挑阳台，从多角度欣赏外面的景观环境。

写字立面简洁，正立面成对称的格局，虚实结合，整体采用灰色色调并灵活运用红色，达到统一中有变化，变化中有统一的效果。

酒店整体采用弧形对称设计，立面简洁，稳健并具有强烈动感，弧线造型柔和亲切而富有韵律，如此精细的细节设计，使建筑更加人性化。

景观设计

该项目设计以自然骨架为支撑打造了自然式山水景观，四周建筑独享内庭院景观，演绎生态绿肺、天然氧吧，为酒会、展览、露天聚会等提供舒适开敞的景观环境。

中轴广场的设立使园区整体增加仪式感，引导人们视线，增强广场区与酒店区、绿肺花园区的延延续性与互通性，并且带动地下商业，与一期演播大厅相呼应。

Profile

Beijing Dongyi International Media Park is located in an extenstion of Beijing Changan Street and situated in the core of media corridor from Beijing CBD to Ding Fu Zhuang. Its east is neighboring with Fifth Ring Road; west is facing Xing Long Park; north is adjacent to Chaoyang Park; and south is leading to Jingdong Expressway. The buildings of mainly single multi—storey constructes in this park are arranged in an orderly way .Its whole style is simple atmosphere. In addition, the buildings are also endowed to promote cultural ambience. In the aspect of landscape, the park emphasizes on building a natural landscape and a home style office environment. So as to build up a real OFFICE PARK.

Planning and Layout

The park is divided into four parts by "H" shape

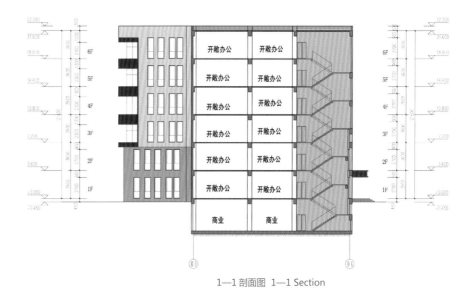

1—1 剖面图 1—1 Section

path. There is a parking lot and two office buildings in the north, a commercial and office complex assigned in the centre of west, high—rise office buildings and business centers as the leading of the park designed in the centre of east, and five single office buildings placed in the south. Buildings in each plot are unique and closely dependent. Among the buildings in this area, the high—rise buildings and business center in the east surrounding each other brings the whole architectural complex together.

Architectural Design

In this project planning, we refer to the spatial layout of Beijing traditional quadrangle then extract local architectural space element and underline the concept of 'courtyard'. Considering the communication among people, buildings and surroundings, we reserve space for it in the initial planning and design, meanwhile we also leave room for post landscape design.

Buildings adopt the modern concise style and echo the colors of the first phase complex by applying classical colors of Chinese style ,namely dark grey, light grey and red. The open—style office space inside the building, the gazebo on the top floor and the projected balcony form a multi—perspective view to enjoy the scenery. The facades of office buildings is concise and their front elevations are symmetrical pattern, consequently obtaining the virtual and real coordination. And the building is using a shape of grey stone and red is also flexibly applied to it. By doing so, the buildings is achieving the effect of unification in changes and changes in unification.

The hotel adopts arc symmetrical design in the whole, and its facade is simple, solid and with dynamic. The arc shape of the hotel is soft, tender and full of rhythm. Through such delicate detailed design, the building is made to be more humanized.

Landscape Design

The project design is a using nature framework as the support to establish the natural scenery .And the inner landscape is exclusive for surrounded buildings, which serves as an ecological green lung and a natural oxygen bar and provides a comfortable and open landscape environment for cocktail parties, exhibitions and outdoor parties etc.

The establishment of axis square adds a ritual sense for the whole park, leads people's view, enhancesthe continuity and interoperability among the square area, the hotel zone and the green lung garden sector, drives underground commerce and echoes the concert hall of the first phase.

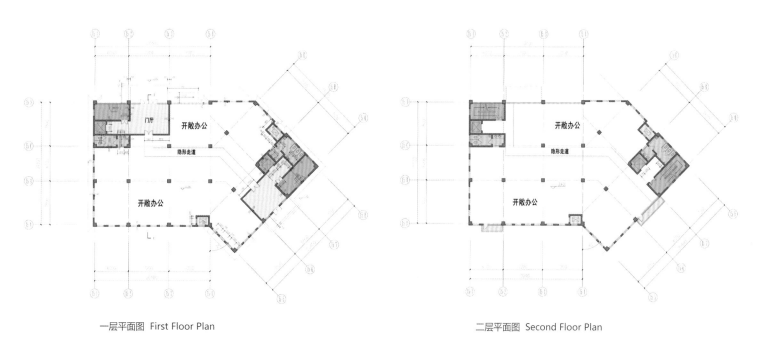

一层平面图 First Floor Plan

二层平面图 Second Floor Plan

NEW CHARACTERISTICS | 新特色

ELEGANT, NOBLE, ROMANTIC FRENCH CLASSIC COMMUNITY

| Oriental Red County

优雅 高贵 浪漫的法式风尚社区 —— 东方红郡

项目地点：中国内蒙古自治区呼和浩特市
开 发 商：内蒙古东方房地产开发有限公司
建筑设计：北京寰亚国际建筑设计有限公司
设 计 师：赵士超、王金山、苏凯、蔡虎、刘彤彤
建筑面积：1 092 200 m²
容 积 率：2.1
绿 化 率：32.98%

Location: Hohhot Innner Mongolia, China
Developer: Inner Mongolia Oriental Real Estate Development Co., Ltd.
Architectural Design: Beijing Huanya International Architecture Design Co., Ltd.
Designers: Zhao Shichao, Wang Jinshan, Su Kai, Cai Hu, Liu Tongtong
Floor Area: 1,092,200 m²
Plot Ratio: 2.1
Greening Ratio: 32.98%

项目概况

　　项目坐落于呼和浩特市金川工业园，总规划用地面积454 830 m^2。基地北侧紧邻金海大道，是金川区通往呼市老城区主要道路，西侧与南侧为城市规划道路。本项目属新城区开发用地，地理位置和市政交通较为优越，距伊利工业园3 km，处在新区核心位置。

规划布局

　　项目总体规划思路：法式风尚社区，尊贵综合大盘。在项目总体规划上，将本地块从东到西划分成A\B+C\D三个主体，三个地块间由一条30 m宽中心景观轴相连，在社区内设计两条规划主干道，贯穿地块南北。

　　A地块以环行道路形式设计，在道路内侧设计以多层和小高层为主的景观式住宅，在两侧建筑围和处，设计小区景观中心。因项目北侧比邻主要道路，考虑到城市主干道对本地块的影响，设计时在地块北侧设计沿街商业街。商业街以菱形分隔为整体布局，在商业街的东

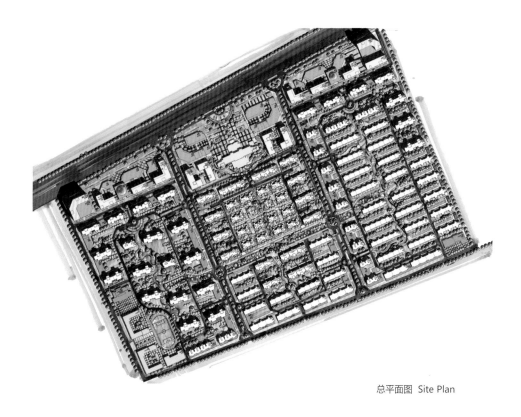

总平面图 Site Plan

NEW CHARACTERISTICS | 新特色

侧设计综合性商业广场，在步行街上设计17层LOFT公寓。

B地块定位为高端住宅区，以联排别墅和花园洋房组成。联排别墅和洋房通过南北两栋楼之间围和，形成景观花园。联排别墅区采用半地下停车库方式，洋房采用地下停车。在B地块的北侧设计社区中心会所，会所建筑面积12 000 m²，服务于整个地块，沿会所的两侧设计LOFT公寓和商业步行街。

C地块以舒适型住宅为主，由多层花园洋房和多层住宅组成，建筑间通过拉大间距，形成户户均能观景的布局。

D地块以高层住宅和教育用地为主，建筑分布呈点状式分布，在地块左下角为教育用地，设置小学、中学和幼儿园。

建筑设计

建筑设计风格为法式建筑风格，它介于新古典主义和现代主义之间，强调建筑物的高耸、挺拔，给人以拔地而起、傲然屹立的非凡气势。通过建筑外立面设

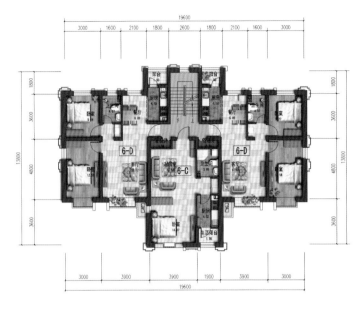

标准层平面图 1　Plan for Standard Floor 1

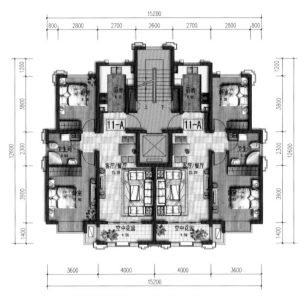

标准层平面图 2　Plan for Standard Floor 2

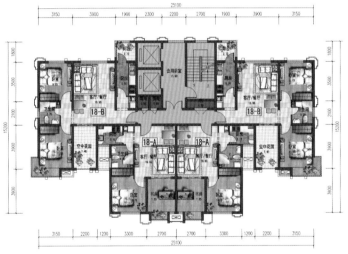

标准层平面图 3　Plan for Standard Floor 3

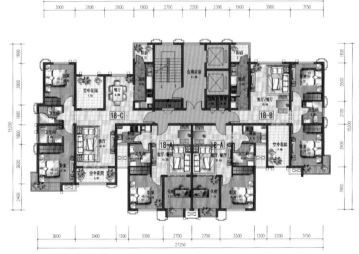

标准层平面图 4　Plan for Standard Floor 4

计，凸显本项目的综合竞争力和城市形象感染力。

小区建筑气质典雅高贵，体型错落有致，注重细节处理。法式建筑是经典的，而不是时尚的，是经过数百年的历史筛选和时光打磨留存下来的。它是一种基于对理想情景的考虑，追求建筑的诗意、诗境，力求在气质上给人深度的感染。风格偏于庄重大方，采用对称造型，营造恢宏的气势、豪华舒适的居住空间，屋顶采用孟莎式，坡度有转折，上部平缓，下部陡直，呈现出优雅·高贵·浪漫的风格。

户型设计

户型设计方面，注重居住空间的舒适性，强调居住的品质感。结合不断提高的现代生活方式，通过厅室厨卫的合理配比，达到户型面积的可用性增加，更好的体现了住宅的舒适性、功能性、合理性、私密性、美观性和经济性。

建筑商业、公寓平面的设计，摒弃了以往死板的排布，在保证各功能房间实用的前提下，灵活布置。通过局部小空间收放自如的处理，形成良好的商业流线和适宜的居住空间，为消费者打造真正实用的居住、购物、休闲和娱乐空间……

景观设计

本项目以贯穿东西的中心景观轴为主轴线，沿主轴线往两侧分布，形成户户均能观景的理想状态。在每个建筑组团处设计景观广场，在景观广场处增加健身设施，满足居民休闲和活动要求。社区由中央景观步道从东到西相互联通，加之景观环岛和水景广场的点缀，动与静相互结合。提供给人们一个能够放松心情的自然环境。

NEW CHARACTERISTICS | 新特色

Profile

Located at Hohhot Jinchuan Industrial Park, the project covers an area of 454,830 m^2. It is on the south to Jinhai Road which is the main road leading to Hohhot old town, and on the east and north to urban planning road. It is a part of new urban development land, 3 km away from Yili Industrial Park, boasting a superior geographical location and municipal transportation.

Planning and Layout

Overall planning ideas: noble French classic community. The site will be divided into three parts, i.e., A, B+C, D, which is linked up by a 30m wide central landscape axis. In addition, two main roads were planned running across the whole community.

Ring road encircles Lot A, in which there are mainly multi—storey and small high—rise residences, landscape center locates where they surround, so each residence may enjoy the view. A diamond commercial street lies in the north to protect the residence from the negative influence by the urban main road. A comprehensive commercial plaza situates to the east of the commercial street and a

17—storey apartment stands on the street.

Lot B is high—end residential area, comprised of townhouses and garden houses. Both of them enjoy a garden view. Townhouses are provided with semi—underground parking place, while garden house are with underground parking place. On the north the lot, there is a 12,000m² club with LOFT apartments and commercial pedestrian street on both sides.

Lot C is for comfortable multi—storey garden houses and multi—storey residence, large distance between these buildings ensures the view for each residence.

Lot D is for high—rise residence and education facilities. Dotted buildings spread all over. Primary school, middle school and kindergarten are arranged in the lower left corner of the lot.

Architectural Design

French architectural style ranges between neoclassical and modernist, which emphasis on towering, tall and straight, giving the extraordinary momentum to where they stand. Through the building facade design, the project's overall competitiveness and image of the city was highlighted.

Buildings in the community are elegant with a well—proportioned figure and refined detail process. The French architecture is classic rather than stylish, survived after hundreds of years of history. It is based on the ideal scenario, the pursuit of poetic of poetry, strives to express the deep infection. They are somewhat dignified and generous, using symmetrical shape to create a magnificent momentum and luxurious living space, especially the mansard roof, showing an elegant, noble and romantic style.

House Layout Design

Great attention is paid to the comfort and quality of living space. Inspired by improving modern life style, designers set reasonable proportion of the interior to increase efficient usage of inner space, to better reflect the residential comfort, functionality, rationality, privacy, aesthetics and economics.

Commercial space and apartment abandoned the previous rigid arrangement, arranging flexibly in premise of ensuring practical. A good business flow line and a suitable living space for consumers were formed, thus to create a truly practical living, shopping, leisure and entertainment space for the customers.

Landscape Design

Central landscape axis runs from the east to the west, to form an ideal state that each residence may enjoy the landscape. Landscape plaza is designed for each group, in which there are fitness facilities to meet residents' leisure and activity requirements. East—west central scenic path is decorated with roundabout landscape and waterspace plaza, providing natural environment for the residences to relax themselves.

CONVERGING CHARACTERISTICS OF URBAN CORE DISTRICTS

| Huacai International Apartment

汇聚都市核心区特征的高品质住区 —— 华彩国际公寓

项目地点：中国北京市
建筑设计：北京奥思得建筑设计有限公司
占地面积：29 623 m²
容 积 率：3.56
建筑密度：20.7%
绿 化 率：35%

Location: Beijing, China
Architectural Design: BHAD
Land Area: 29,623 m²
Plot Ratio: 3.56
Building Density: 20.7%
Greening Ratio: 35%

项目概况与规划布局

项目位于北京市朝阳区东三环中路，总体规划采用南北两排住宅，中心集中布置绿地的规划布局，并结合局部架空的设计手法增大了外部视觉空间，以缓解绿化纵深小的矛盾。运用都市综合体的设计手法，结合商业中心、步行商业街、以及户外广场、绿地，创造具有都市核心区特点的功能组合，空间尺度、建筑形象和环境特点。实现城市机能和都市形象的更新和发展，同时为开发商创造最有效的商业利益。沿社区北侧住宅设置底层商业设施，与会所及北侧商业中心构成相对集中的外围商业服务设施，住宅的私密性得到体现。

建筑设计

采用短板和点式住宅相结合的设计手法，既保证了居住产品的先进性和市场竞争力，同时增加了社区的通风效果和视觉通透性。首层局部架空，地下车库采光口等设计手法丰富了社区立体空间效果，室内外环境相互融合、生动丰富。南侧以点式住宅为主，增加了南北通透性，大大改善了中央绿地的日照和使用效果。

户型设计

户型方面包括90～110 m²（二室二厅二卫）、140～160 m²（三室二厅二卫）、190～210 m²（四室二厅三卫）三种套型，其中140～160 m²的舒适型住宅为主力户型。

景观设计

景观方面规划运用中国传统园林和日式园林等东方园林设计手法，以取得曲径通幽，小中见大的环境效果。

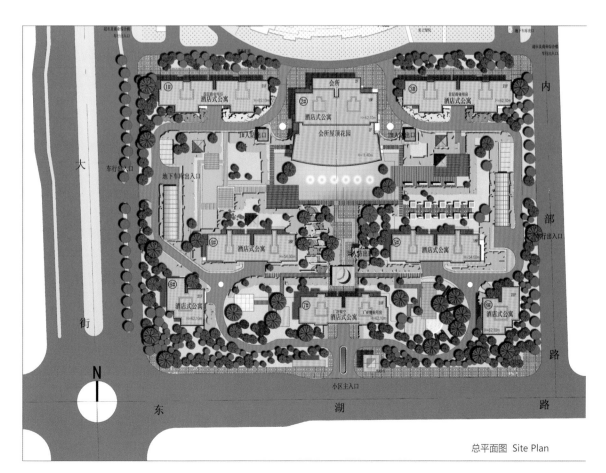

总平面图 Site Plan

NEW CHARACTERISTICS | 新特色

Profile and Design

The project is located in the East Third Ring Road, Chaoyang District. Two rows of residences, north one and south one, embraced green space in the middle, and combined with partial overhead design technique increases the external visual space, to ease the greening depth contradictions. Use of municipal complex design techniques combined with the commercial center, commercial pedestrian street, as well as outdoor square, green spaces, creating a combination of features with the characteristics of the urban core area, the spatial scale architectural image and environmental characteristics. Update and develop the city function and city image, and at the same time for developers to create the most effective commercial interests. Commercial facilities were arranged at the ground floor of north residence, making up the relatively concentrated periphery commercial service together with the club and north commercial center, which ensured the privacy of the residence.

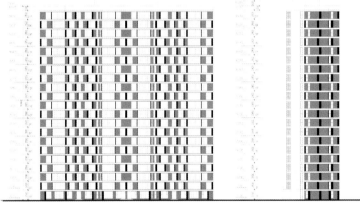

南立面图 South Elevation　　西立面图 West Elevation

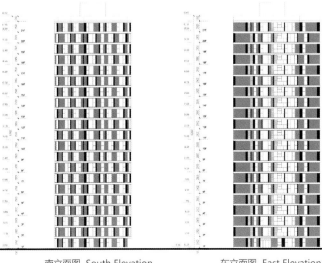

南立面图 South Elevation　　东立面图 East Elevation

NEW CHARACTERISTICS | 新特色

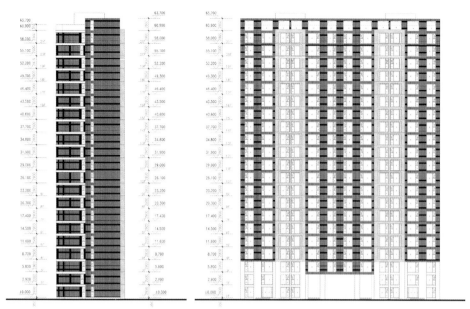

西立面图 West Elevation 南立面图 South Elevation

Architectural Design

Short plate residence combined with point block apartment, not only ensured the advancement and market competiveness of their own, but also increased the ventilation effect and visual permeability of the community. The first floor partial overhead, daylight opening of the underground garage enriched the dimensional spatial effects, merging indoor and outdoor environments, seemed vivid and rich. Point block apartments are mainly in the south, which increased the north—south permeability, greatly improved the sunshine and use of the central green space.

Layout Design

There are three types of houses: 90~110m² (two bedrooms, two living rooms and two bathrooms), 140~160m² (three bedrooms, two living rooms and two bathrooms), 190~210m² (four bedrooms, two living rooms and three bathrooms), in which the one of 140~160m² is the main type.

Landscape Design

In terms of landscape planning, designers adopt the techniques that used to create traditional Chinese garden and Japanese garden, to get the meandering paths, feeling the whole environment step by step.

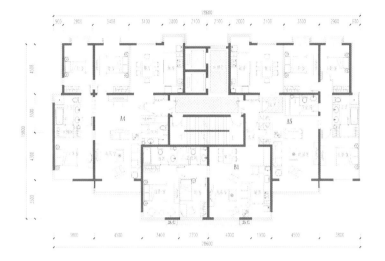

1# 楼户型平面图
1# Building Unit Plan

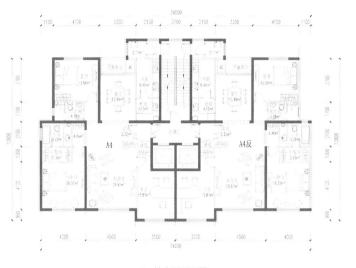

7# 楼户型平面图
7# Building Unit Plan

NEW CHARACTERISTICS | 新特色

THE MAXIMUM OF LANDSCAPE VALUE

| The Planning and Design Proposal of Dwelling District in Sun Island, Nanning

景观价值最大化运动——南宁太阳岛居住区规划设计方案

项目地点：中国广西壮族民族自治区南宁市
项目策划、空间规划、景观规划、建筑设计：广州市金冕建筑设计有限公司
总用地面积：153 871 m²
总建筑面积：402 691 m²
容 积 率：2.13
建筑密度：21.1 %
绿 化 率：36.7 %

Location: Nanning, Guangxi, China
Project Planning, Space Planning, Landscape Design,
Architectural Design: King Made Group, Guangzhou
Total Land Area: 153,871 m²
Total Floor Area: 402,691 m²
Plot Ratio: 2.13
Building Density: 21.1 %
Greening Ratio: 36.7 %

项目概况

项目处于南宁市中心片区北侧的城郊结合部，南临城市二环线，北连罗伞岭水库，东接南宁市人民医院，西侧现状为生产型绿地。地块狭长，沿水库边界线长达1 000 m。项目交通优越，空气清新，景观优美，为高品质的住宅小区提供了良好的外部环境。

规划布局及空间组织

动静分区，合理布局——合理的动静分区，减少人流对小区内部住宅组团的干扰，提高了内部住宅组团的居住品质。

动——把步行广场、商业中心、幼儿园和酒店设置靠近主入口一侧，方便居民上下班接送小孩、购物、社交、人群集散等。东侧另开一条通往酒店的路，在不影响住宅交通的同时也方便酒店客人的生活。

静——在城市二环线旁，利用商业与板式高层把外面的噪音和人流进行阻隔，使小区内部组团的私密性、领域感、归属感得到进一步的提升。

曲线形住宅组团，最大化了水体景观——曲线形组团设计将1000 m的水体景观界面扩大到了1350 m，使90%以上的住户都能享受到水体景观。

点式布局与别墅的完美穿插——15万 m² 用地面积，76套一线湖景别墅，2.13的容积率。

点式空中别墅独立于水体切割而成绿水环绕的小岛上，与迂回环绕的联排别墅形成巧妙穿插。通过高低搭配最大限度的获取水体景观和开阔的视野，提高产品的景观附加值。

每户别墅设有独立的车库和私家花园，同时高层别墅也设有空中花园，它们在最大限度获取景观的同时，自身也成为美丽景观中的一部分。

景观设计

水体景观的高效利用

①：景观主要是以水为特色，利用水库的自然水体景观，将水引入小区；

②：在水边设置亲水平台和小型广场，满足小区居民休息、休闲、社交等一系列功能；

③：地块两端设置码头，方便小区另一端居民通过游艇来回穿梭，不但解决了小区核心服务区服务半径的问题，还让居民在回家的路上增加了趣味性。

通透的景观视线——项目内部水库可视界面最宽可达780m，最窄也有120m。点式空中别墅沿水库边界线穿插布置，使其最高可拥有140°无阻挡水体景观视线。

贯穿小区的步行栈道——1260 m的滨湖休闲步行栈道根据水岸线的变化而变化，将各个小岛串联起来，步移景异，同时水体的若隐若现形成一种曲径幽深、柳暗花明的景观特色。

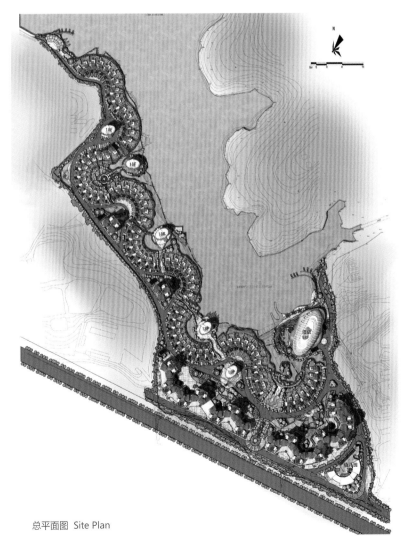

总平面图 Site Plan

Profile

The project is located in the outskirts, north side of the central Nanning with Second Ring Road to its south, Luosanling Reservoir to its north, Nanning People's Hospital to its east and production—oriented greening to its west. The plot is in long and narrow shape, with a length of 1,000 m along the reservoir boundary. It occupies a convenient traffic and fresh air, graceful landscape, offering great outdoor environment to high—quality dwelling districts.

Planning & Layout and Spatial Organization

Reasonable Layout with static and dynamic zoning: reasonable static and dynamic zoning will lower the interference from crowds in the residential groups inside, improve the living quality.

1. Dynamic: Pedestrian Plaza, Commerce Center, a kindergarten and a hotel are positioned close to the main entrance, available for parents' picking kids up, shopping, social contact, crowd collection and distribution etc. Another road to the hotel is set in the east side to guarantee the convenient living in the hotel without interference in the residential traffic.

2. Static: the commerce and slab high—rises are designed to cut off the noise and crowds outside next to the Second Ring Road, promoting the privacy, field sense, belonging sense in the interior residential groups.

Curved—shaped residential groups maximizing the waterscape: a curved—shape group design expands the 1,000 m waterscape surface to be 1,350 m, making more than 90% residents able to enjoy the waterscape.

Point—type layout perfectly interwoven with the villas: a land area of 150,000 m^2, 76 suits villas with first—line lake views and a plot ratio of 2.13.

The hanging villas are located in an island surrounded and isolated by water, showing an ingenious interweave with the surrounding townhouses. The design has maximized the waterscape and the open vision filed, enhanced the landscape added value by the height variation.

Each villa owns the private garage and garden, and the high—rising villas have hanging gardens, which maximizes the landscape meanwhile making themselves parts of the graceful landscape.

Landscape Design

High efficiency in using waterscape: as a major landscape feature, water is led into the community with the adoption of natural waterscape of the reservoir. Water platforms and small squares are located near water for people's leisure, recreation and social contact etc. Wharfs are set in both side of the plot for the residents in the other side to come and go by yachts, which not only solve the service radius problem in the coral service area, but also take the fun in people's way back home.

Unobstructed landscape views: the visual interface width of the interior reservoir is 780m at maximum and 120m at minimum. Point—type hanging villas are positioned along the reservoir boundary, which guarantees the 140° unobstructed waterscape views at the top point.

Walking path throughout the district: the 1,260m leisure walking path at the lakeside varies with the changes of shorelines. Connecting the isles with the looming water, it offers a landscape featuring a maze—style vista.

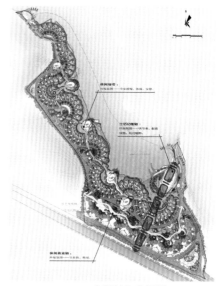

空间结构分析图
Analysis of Space Structure

景观分析图 Landscape Analysis

PASSIONATE, ENERGETIC, SIMPLE, ELEGANT, BRIGHT

| Vanke Summer Showflat

激情 活力 简约 清新 明亮 —— 万科缤纷夏日样板房

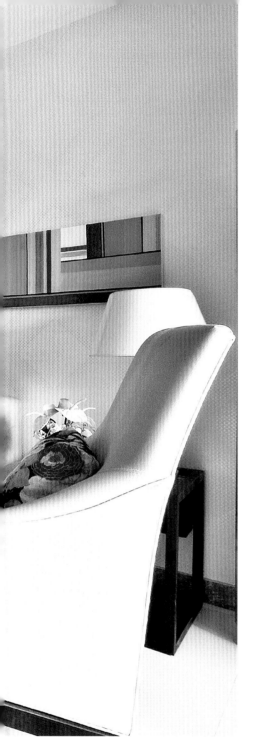

项目地点：中国广东省佛山市
室内设计：广州共生形态工程设计有限公司/
COCOPRO.CN

Location: Foshan, Guangdong, China
Interior Design: C&C Design Co., Ltd./COCOPRO.CN

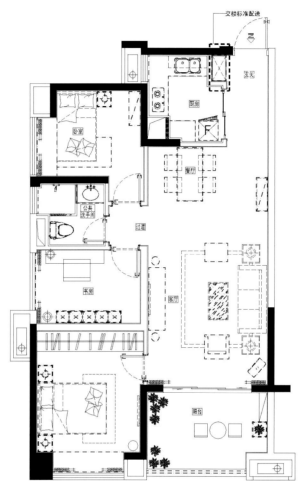

标准层平面图 Plan for Standard Floor

NEW SPACE | 新空间

缤纷夏日，都市里简约住宅也要活力四射。该项目客厅错拼的彩色长条装饰画搭配白色简约的沙发，波浪彩条的抱枕，就连蒲公英忍不住将要随风起舞。餐厅嫩嫩草绿色调，雏鸟形的餐巾扣、白色简约的树形烛台完美的搭配，精致明亮的厨房，彩色艳丽厨具，诱人的美食。小孩房小巧的木马，缤纷的泡泡床品，活泼的色彩表现了天真可爱。

Summer is coming, and the residence in city also wants to liven up the summer days. Sitting room is decorated with colorful strip paintings, white elegant sofa and color—stripped pillows. This colorful atmosphere will bring the dandelion the passion to dance with the wind. Dining room is decorated in light green with bird—shaped napkin rings, white elegant tree—shaped candleholder, bright boutique kitchen, colorful cookers and delicious food. Kids room is furnished with a small hobbyhorse and colorful bubble beddings to match the innocence and loveliness of children.

NEW IDEA | 新创意

CREATIVE LIVING SPACE WITH WIDE VISION, FULL OF VIGOR
Guangzhou Science & Technology Staff Living Quarter

视野开阔、充满活力的创意性居住空间——广州科学城科技人员集合住宅

NEW IDEA | 新创意

项目地点：中国广东省广州市
建筑/景观设计：日本佐藤総合建筑设计事务所
　　　　　　　广东省建筑设计研究所
占地面积：39 957 m²
建筑面积：105 407 m²
绿 化 率：40.9%

Location: Guangzhou, China
Architectural/ Landscape Design: AXS Satow Inc.
The Architectural Design & Research Institute of Guangdong Province
Land Area: 39,957 m²
Total Floor Area: 105,407 m²
Greening Ratio: 40.9%

总平面图 Site Plan

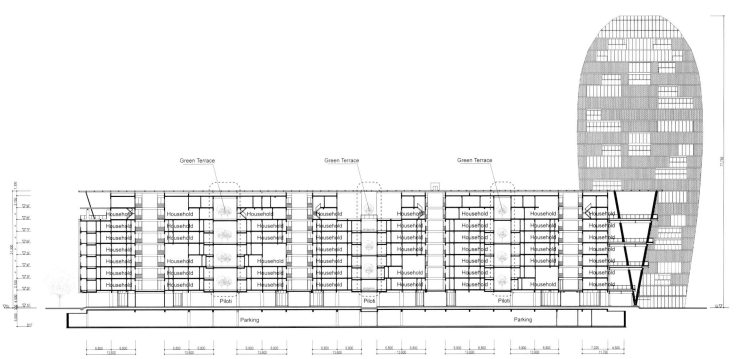

剖面图 Sectional Drawing

NEW IDEA | 新创意

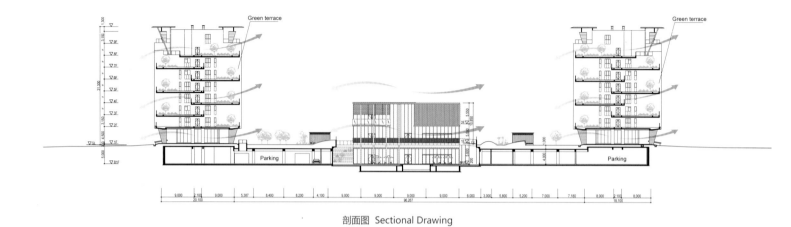

剖面图 Sectional Drawing

项目概况

该项目位于广州科学城地块区域，是为在科技园工作的海外技术人员提供住宅的项目。其周边具有高质量的生态环境，完善的城市基础设施，交通便捷，同时能够很好的满足科学城高级技术人员在此工作与生活的需求。

规划布局及建筑设计

建筑利用当地盛行风的特点有效地避免热带风暴及强降雨对住户造成太大影响。四个低层单元的连接处有一片空地，其中包括一条通向一层的风道。风道作为居住空间的延伸也能够帮助住户抵御酷热的天气以及提供避雨的半开放空间。分层的露台作为连接各个居住单元的开放空间方便邻居之间的沟通和联系。

由于园区绿化面积大，住户拥有宽阔的可视域并增添了建筑的活力。每个单元内的天窗阻隔了太阳光并作为屏风挡住住户的视野，但由于其透明的颜色仍不影响视野范围。每栋建筑六楼都设置了摘星阁，其功能类似于天窗具备良好的通风效果。

NEW IDEA | 新创意

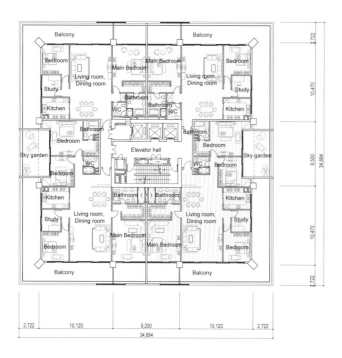

十六层平面图 16st Floor Plan

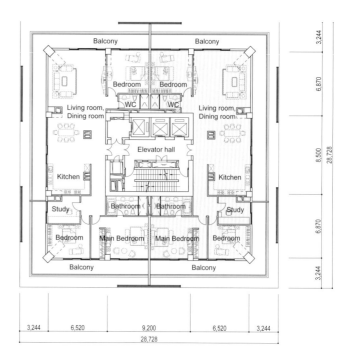

二十二层平面图 22nd Floor Plan

Profile
This is residential building for the scientists working projects overseas in order to develop Science Park. Its surroundings with high—quality ecological environment, perfect urban infrastructure, convenient transportation, and can well satisfied those scientists living need.

Planning Layout and Building Design
It has the function of using prevailing wind as regional characteristic to prevent the subtropical heat and rain. At the connection between the four lower units there is void space which obtains the wind path in first floor. This 'wind path' works the mutual space as the extension of living space and also controls severe heat and rain as semi—outdoor. The layered green—terraces are open spaces between each unit which help residents to communicate neighbors.

Because the green comes into the first floor, it gives extra visual sense of unity with the surrounding green space and useful essence as the activity along the building. The vertical louver in each unit controls light from sun and also works as a screen to block people's eye but still has high transparency. The sky court appeared in the center of elevation of residential tower is located at every six floors and functions as green window in order to let a wind inside from outside.

UNITY OF DWELLING SPACE WITH SINGLE—SLOPING ROOF DESIGN
Can Bisa House

单一倾斜屋顶设计统一住宅空间—— Can Bisa住宅

项目地点：西班牙巴塞罗那
建筑设计：Batlle i Roig, arquitectes
摄　　影：A. Flajszer

Location: Barcelona, Spain
Architect: Batlle i Roig, arquitectes
Photography: A. Flajszer

项目概况

Can Bisa 是一座19世纪末的公寓，现为Vilassar de Mar所有。它位于Riera de Cabrils 河道附近的一个街区，那里曾经是一座工厂，现在已经被拆毁。由于它的历史文物价值以及在城镇中所处的战略地位，委员会考虑将其打造成一个理想的文化设施场地，完成一个社会住房建筑群。

规划布局与建筑设计

在该项目规划进行中，建筑师很快了解到Can Bisa不可能容纳所有重要的用途功能，项目建筑基本是交付委员会办公室使用，因此礼堂、酒吧和会议室被设置在住宅建筑的地面楼层。

住宅体块的总体规划将其分为两个具有不同高度的体量，形成了一个"L"形，尊重和保证了街道的宽度；同时也腾出了一定的空间来设置礼堂。该项目的建造主要是通过单一倾斜屋顶设计将住宅空间统一起来，在街道上设计出连续的外观，同时对室内空间进行合理布局，实现住宅间的相互联系。两座建筑之间的通道就是通往该建筑群的主要入口。建筑的规模更容易让人联想到庭院设计。它由一系列休闲和娱乐空间组成，周围的设施也是为各种各样活动的举办创造了有利条件。艺术家Carme Balada专门设计了陶制花盆，用于装饰该栋单元楼。

住宅体块和公寓都采用了白色灰泥装修，呈现不同宽度的条状纹理。阳台和百叶窗采用金属和格子砖石，旧厂区的一部分被保留成为新建筑的一个立面，向众人展示着它的历史。

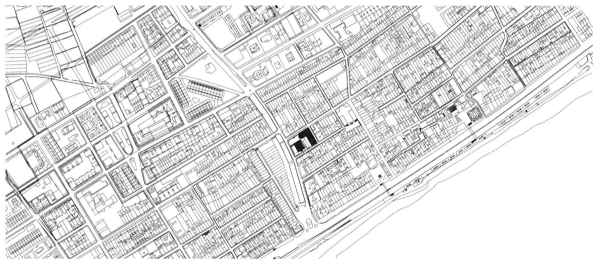

NEW IDEA | 新创意

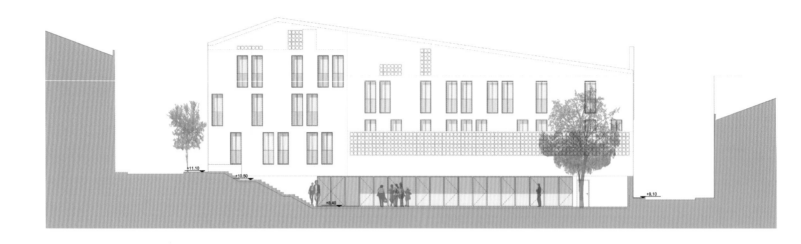

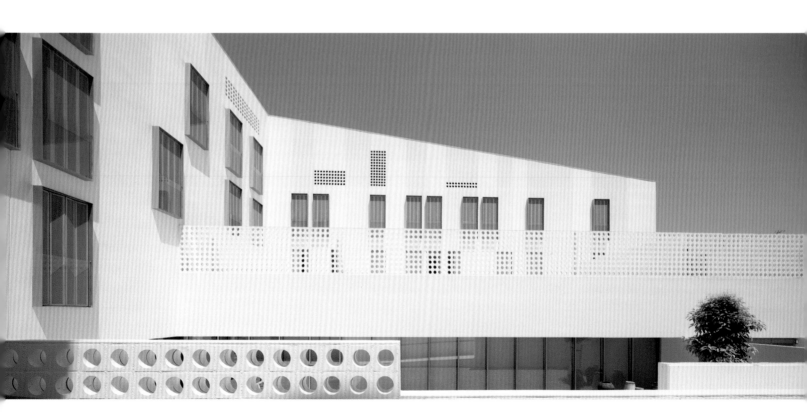

Profile

Can Bisa is a late—19th century mansion now owned by Vilassar de Mar Council. Situated on the Riera de Cabrils watercourse, it occupies part of a street block that used to include a factory, now demolished. Its historic and heritage value and the strategic position it occupies in the town as a whole led the Council to consider it the ideal venue for a cultural facility, completing the complex with a social housing block.

Planning Layout and Architectural Design

During the implementation of the facility's programme it rapidly became apparent that it would be impossible to accommodate all the necessary uses in Can Bisa, which is basically given over to council offices; the auditorium, bar and meeting rooms are therefore arranged on the ground floor of the residential building.

For the dwellings, the General Plan specified two volumes of different heights, joined to form an L—shape that respects the width of the streets and mounted on a larger base which is envisaged to accommodate the auditorium. The aim of the project was to unify the volumes by means of a single sloping roof, construct a continuous facade onto the streets, leading to the dwellings, and lay out the interior volume to encourage interrelation with the mansion. The passage generated between the two buildings becomes the main entrance to the complex. Its dimensions are more suggestive of a courtyard and, together with the gardens around Can Bisa, it comprises a series of leisure and recreation spaces that complement the activities offered by the facility. Ceramic planters designed specially by the artist Carme Balada serve to organize the complex.

Both the housing block and the mansion are given a textured finish of white stucco, in strips of varying widths. The balconies and shutters are made of metal, with masonry lattices. Part of the old factory is retained in one of the fa ades to bear witness to its past.

NEW IDEA | 新创意

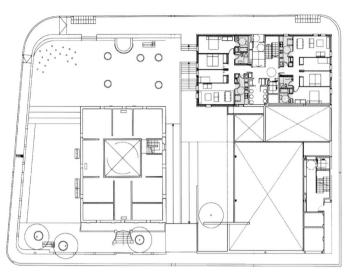

PLANTA BAIXA
GROUND FLOOR PLAN

PLANTA SOTERRANI
UNDERGROUND FLOOR PLAN

COMME[RCIAL]
BUILDINGS

RCIAL 商业地产

P130
TARSU购物娱乐中心：
交错重叠的多功能立体商业空间

P138
山西天美新天地：
三位一体的高端城市综合体

P144
星月坊：
定位于未来的多功能商业项目

STAGGERED MULTI—FUNCTIONAL COMMERCIAL SPACE

TARSU SHOPPING & ENTERTAINMENT CENTER

交错重叠的多功能立体商业空间—— TARSU购物娱乐中心

项目地点：土耳其大数市
业　　主：科里奥
建筑设计：Yazgan Design Architecture
总建筑面积：63 000 m²
摄　　影：Yunus Özkazanç、Kerem Yazgan

Location: Tarsus, Turkey
Employer: Corio
Architectural Design: Yazgan Design Architecture
Total Floor Area: 63,000 m²
Photography: Yunus Özkazanç — 2012, Kerem Yazgan — 2011

项目概况

　　Tarsu是一个建筑面积为63 000 m^{2}的购物和娱乐中心,坐落在土耳其大数市。含一个地下室的两层楼建筑,拥有大型购物和娱乐设备,如大型综合超市、技术厅、娱乐中心和电影院,它拥有75家店铺是大数市最大的购物中心。该建筑一楼延伸到地下,这使它在传统的大数街道上体形突出。超大的体形和灯箱以独特的三维相互重叠,与该市不同的历史、文化和社会环境,一道丰富了这座城市。

规划布局

　　在大数市水显得十分重要,影响着大数的历史进程。该城始建于港口城市的交汇点,处于倾斜的托罗斯山脉与河流之间。水孕育着生命、生活和城市的一切;自然、文化和水在整个城市的历史交织在一起。命名为"Tarsu"是因为它的发音中有"水",土耳其语里"su"的意思是"水"。"Tarsu"表示河水在这里相遇的意思。因此,水元素是无处不在,建筑物的外侧和内侧有九个池子和三个喷泉。在二楼的美食广场规划有一个巨大的水族馆,接待游客。

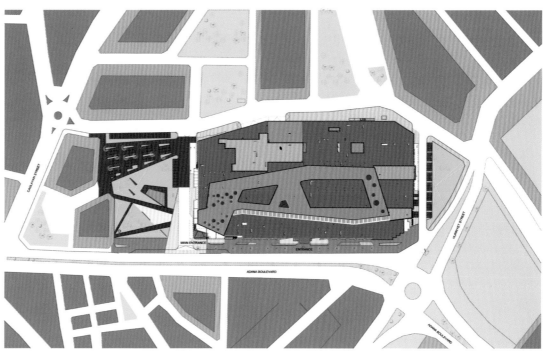

总平面图 Site Plan

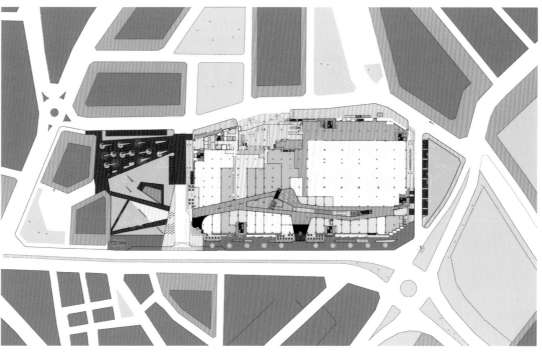

地面层平面图 Ground Floor Plan

COMMERCIAL BUILDINGS | 商业地产

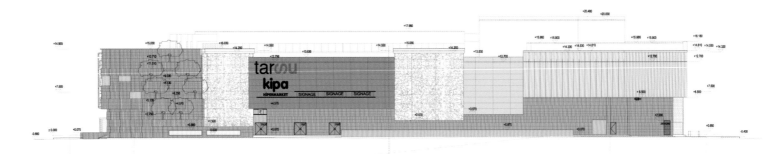

东立面图 East Elevation

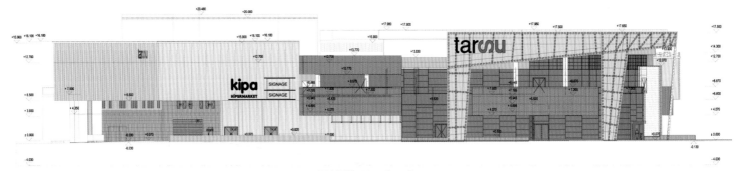

西立面图 West Elevation

建筑设计

该建筑的设计，既使用技术手段，也利用自然通风。根据自然通风和采光标准要求而设计的屋顶的流线型窗户。白天，自然通风时风门自动关闭，室内感应探头感应到氧气减少。到了晚上，这些风门会自动打开，以提供室内及店内冷气。商店的业主，当他们早上来，并不需要使用额外的能量用于通风。这就能保证室内凉爽，节省能源。

由于没有安装过滤设备和管道系统的池子，需要每周补给。Tarsu 采用电子过滤和循环水补充系统，一个循环需要3个月，而不是每周。这使得水消耗在能够大大减少。池外建筑安装有热和风量感应探头，使水的流失减少到最低。

项目采用了不同的材料，它们有玻璃、砂岩、玛瑙、铝复合板和铝板等。在夜间莹光灯的照射下，随时间变化着颜色，辉映着这座多彩的城市。

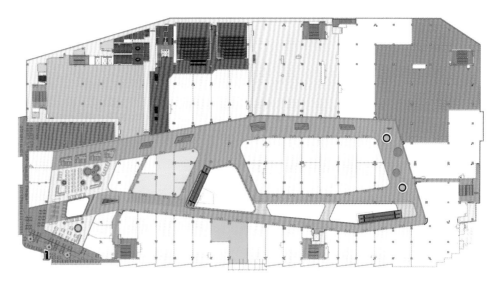

一层平面图 First Floor Plan

COMMERCIAL BUILDINGS | 商业地产

Profile

Tarsu is a 63,000 m² of shopping and entertainment center, which is located in Tarsus, Turkey. The two—story building with a basement floor has 75 shops with large scaled shopping and entertainment units, such as, a hypermarket, a technology shop, an entertainment center and cinemas, making it the largest shopping center in the city. The building's first floor volume extends over the one belonging to the ground floor, referring to the protrusion of upper volume to the street at traditional Tarsu houses. These extended volumes and light boxes overlap with each other with a peculiar three dimensional dynamism, addressing the overlapping of diverse histories, cultures and social environment that make up of the unique richness of the city.

Planning and Layout

Water is a prominent element affecting the history of Tarsus. The city was founded on the meeting point of city harbor with the river that flows between Toros Mountains that the city reclined upon geographically. Water brings life, trading and soul to the city. Nature, culture and water are intertwined throughout the city's history. The name "Tarsu" is selected for it's wordily allusion to "water"; hence, "su" means "water" in Turkish. "Tarsu" denotes the meeting of Tarsus city with water. Hence, water elements are everywhere; there are nine pools and three fountains at the outside and the inside of the building. A massive aquarium meets the visitor at the food court of the upper floor.

Architectural Design

The building's design, coupling with it's technology, enable the utilization of natural ventilation as well. Linear windows at the roof are designed according to standard requirements for natural ventilation and daylighting. Natural ventilation dampers automatically close when their censors detect a decrease in the interior oxygen quality. At night, these dampers automatically open to provide cooling of the interiors, including that of shops. The shop owners do not need to utilize extra energy for ventilation when they come in the morning. That condition decreases the energy used for cooling of the

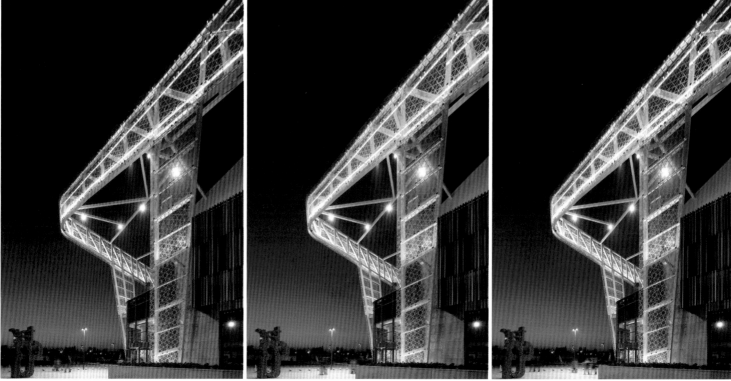

COMMERCIAL BUILDINGS | 商业地产

interiors.

Pools that do not have filtration and circulation systems require water replenishment every week. Tarsu pools have electronically operated filtration and circulation systems that require water replenishment in 3 month, rather than weekly periods. That enables reduction of water consumption at a significant degree. The pools outside the building operate with heat and wind censors so that the water loss is reduced to minimum.

They are covered with different materials, which are glass, sandstone, onyx, aluminum composite panels and aluminum sheet. They are lit by LED lighting elements at night, with their colors changing at sequential time periods, enhancing the allusion to the colorful palette that make up of the Tarsus urban culture.

TRINITARIAN HIGH—END HOPSCA

Tianmei Century Plaza

三位一体的高端城市综合体—— 山西天美新天地

项目地点：中国山西省太原市
客　　户：山西天美新天地购物中心有限公司
建筑/室内设计：RTKL
建筑面积：50 000 m²
摄　　影：© RTKL/David Whitcomb

Location: Taiyuan, Shanxi, China
Client: ShanXi Taiyuan Tian Mei Ming Dian Shopping Center Co., Ltd.
Architectural/Interior Design: RTKL Associates Inc.
Floor Area: 50,000 m²
Photography: © RTKL/David Whitcomb

项目概况

天美新天地位于山西太原城南第一大街长风街的核心地段,三面临街,在太原市"南迁西进"的城市发展战略下已形成集中央商务区、高档住宅区、市政文化区为一体的城市中心地带。交通极为便利,相邻的五星级酒店、高端写字楼、山西煤炭进出口集团都将为天美新天地带来高端而稳定的客流。

规划布局及建筑设计

项目整体的设计思路是:将城市的文化历史与其现在和未来的成功连在一起。为此,设计师规划了一个 50 000 m² 的流动空间,6楼高的中庭;一个综合图形和标志规划,吸引着购物中心的富裕顾客。

在建筑设计上,RTKL的建筑师面对的挑战是要将已有的三栋大楼连接起来,而室内团队的挑战则是如何实现商场大厅与男性时尚区风格一致。

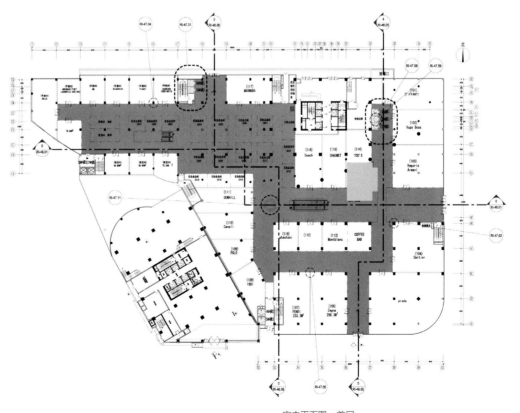

室内平面图—首层
Interior Floor Plan— Ground Floor

COMMERCIAL BUILDINGS | 商业地产

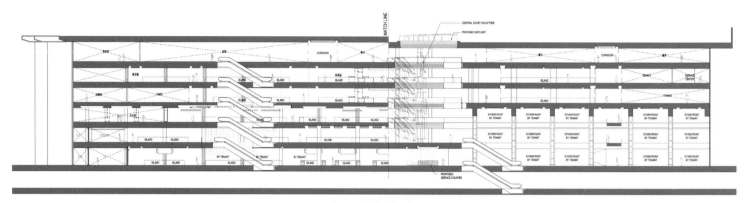

剖面图 A SectionA

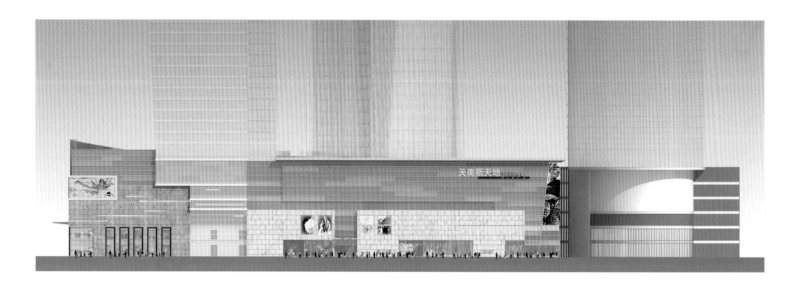

COMMERCIAL BUILDINGS | 商业地产

Profile
RTKL's architects were challenged to tie together three existing buildings, while the interiors team was charged with creating a seamless, open transition from the mall area to the department store, all uniquely targeting men's fashion. After working with the client and deciding on a concept of linking the city's cultural past to its current and future success, the designers created a 50,000—SM flowing space with a six—story atrium and a comprehensive graphics and signage scheme that appeals to the shopping center's affluent clientele.

Planning and Architectural Design
The project locates in the coral plot of Changfeng Avenue, the first biggest avenue in the south Taiyang, Shanxi. With streets on 3 sides, it has become a urban center as central business district, a high—end residence, municipal cultural district, under Taiyuan's urban development strategy of "moving to the south and extension to the west". Apart from the convenient traffic, the adjacent five—star hotel, high—end office buildings, Shanxi Coal Import & Export Group Co., Ltd. (SCIEG) will bring the high—end and stable customers.

MULTIFUNCTIONAL COMMERCIAL PROJECT ORIENTED TO FUTURE

Moon Bay Street
定位于未来的餐饮功能商业项目 —— 星月坊

项目地点：中国江苏省苏州市
开 发 商：中新苏州工业园区置地有限公司
建筑设计：AAI国际建筑师事务所
用地面积：15 000 m²
建筑面积：14 800 m²

Location: Suzhou, Jiangsu, China
Developer: China-Singapore Suzhou Industrial Park Land Co., Ltd
Architectural Design: Allied Architects International
Site Area: 15,000 m²
Floor Area: 14,800 m²

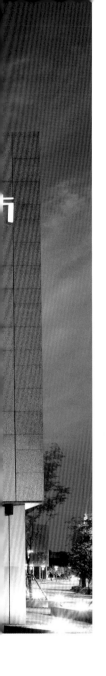

项目概况

项目位于独墅湖畔的月亮湾商务核心区，今后这里将引入大量高档写字楼、宾馆等，以完善科教创新区的办公、展览、文化、休闲等功能，从而带动整个区域基础配套功能的升级优化。

规划布局

在设计之初，建筑师对项目的商业策略进行了较为细致的研究分析。作为区域内率先开发的商业项目，一方面坐拥得天独厚的先发优势，另一方面，其大量客户都潜在于未来，因而如何对其准确定位就显得很重要。原设立目标是作为对街酒店的配套设施，拥有丰富业态，但15 000 m²的面积使其如仅成为配套，自身运营必成难点。仔细分析周边，在近期已有大学城及部分居民区，且月亮湾已成苏州新景观中心——部分市民的休闲去处。从长远而言，周边将出现一群办公、公寓、酒店，潜在一定消费群；就自身而言，用地狭长、面积不大，如以一种主力业态出现，可吸引周边2 000~3 000 m的人群。于是，一个中小型结合、中西合璧的多元化特色餐饮区方案浮出水面，项目设计通过餐饮的群聚效应，不仅最大可能地吸引当前有限的消费群体，树立起自己的品牌形象，同时为今后大量消费群的到来做好准备，并将在极大程度上解决上班族用餐问题。

由此，建筑师给出的设计策略是：

1、将主入口设于东侧，对入月亮湾的主通道、对大学城方向，设置大尺度中型主力店，以连锁店为引入目标，力争以其品牌声望，吸引过往人群注目。同时，结合立面、景观设置招牌，体现整个商业的内容，并引起过客联想。

2、基地中间正对酒店中轴开放广场，既延续了月亮湾总体规划中的向心引力，又将成为吸引酒店住客的次主力。

3、在基地尽端设一个下沉广场，环境安静、景观优良，在此安排较高档的主力餐饮店。如此有节奏地形成三个序列。

相应的，在餐饮模式上，从入口至中间广场，因其临近入口、地段热闹，设置西餐、速食，以满足过

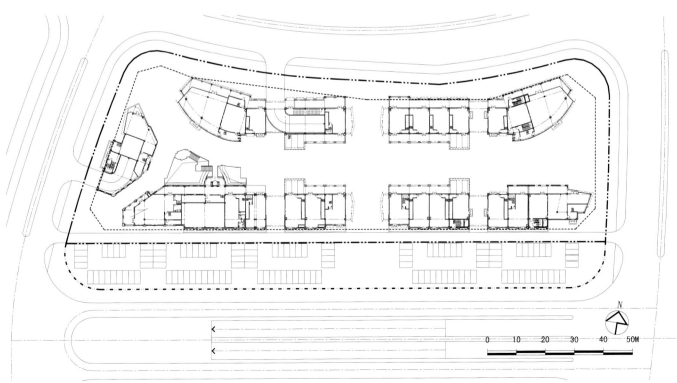

总平面图 Site Plan

COMMERCIAL BUILDINGS | 商业地产

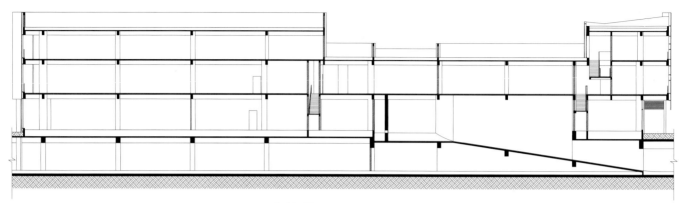

2# 楼 剖面图 Section of 2# Unite

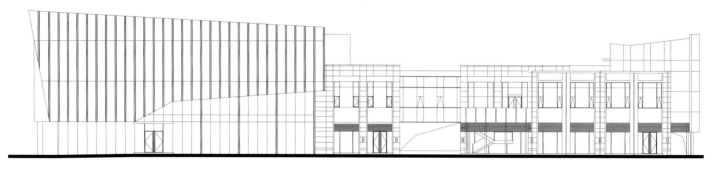

2# 楼 立面图 Elevation of 2# Unite

往及一般酒店和商务人群的需求。而从广场至端头一段，则相对远离人潮、显得幽静，设置中餐、特色餐，以吸引高端商务宴请。基于以上的业态设想，在建筑体量上，呈大——中——大分布，其内部开间和尺寸相对呼应；在空间上，因其狭长，且外部资源不足，放弃低档商铺的想法，而考虑以内为主朝向来聚集人流；从形式上，入口至中段多采用出挑的二层平台，加底层架空，构筑出聚合人气的气氛空间。

建筑设计

建筑以石材、铝板、玻璃幕墙相结合的手法作为标准化的立面基本模块，与整个月亮湾建筑群及对街酒店相呼应。项目局部平面通过前后的进退，形成前后交错的体量，增加了建筑的趣味性；广告牌的设立，成为今后商业自身的特色，也保留了个性发挥的余地；内街中央两侧通过三层的体量，辅以全玻璃幕墙，强化了东西、南北两条轴线的存在感；而下沉广场的彩色玻璃幕墙和景观楼梯，使其跃然成为整个区域的亮点。项目设计结合基地形状，从业态出发合理规划，全力打造重点部位，其余通过控制协调，为今后发挥个性作好了准备。

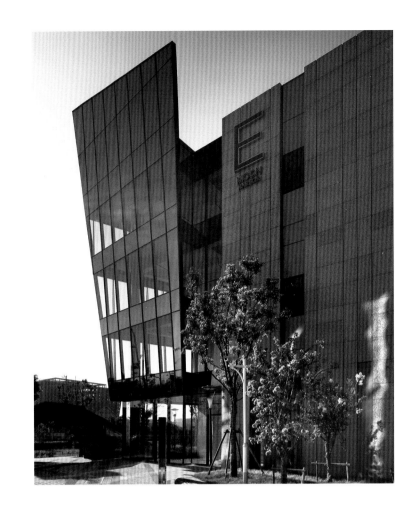

COMMERCIAL BUILDINGS | 商业地产

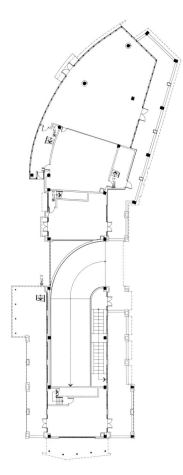

2# 楼 一层平面图 First Floor Plan of 2# Unite

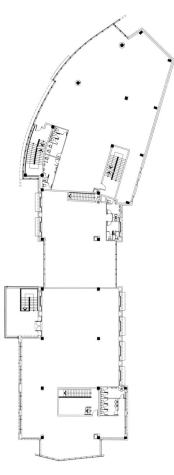

2# 楼 二层平面图 Second Floor Plan of 2# Unite

Profile

This project located in Moon Bay business core area beside Dushu Lake, where will attract a large number of high—end offices, hotels etc, which should improve the function of this area where own the function of office, exhibition, culture, and relaxation purpose, hence drive fundamental assort function to be upgraded and optimized in the whole area.

Planning and Layout

In the beginning of design, designers made careful research and analysis of the business strategy. As the first business project in this region, on one hand, it enjoy privileged first—mover advantage, on the other hand, most of its potential customers are in the future, therefore, how to position its place precisely is rather important. The original aim is to make it as ancillary facility of the across street hotel, possess rich type of business, but owing to its large area of 15,000 m^2, only regard it as ancillary facility will make it difficult to operate. After careful analysis of its surroundings: recently there is university town and some residential area, and Moon Bay becomes a new landscape in Suzhou, as a leisure place for some citizens. Hence, the unique diversification dining area scheme combined

with small and medium size, Chinese and Western elements formed, through cluster effect of catering, the design not only attracted current limited consumer groups, built its own brand image, well prepared for the coming large quantity consumer groups in the future at the same time, but also solved the problem of office worker's meals for the most part.

Thus, architects came up with the following design strategy:
1. Placing its main entrance on the eastern side, over against main channel of Moon Bay and university town, set large—scale medium—sized store, consider chain store as introduction goal to attract passer's eyes. Meanwhile, combine with facade, landscape to embody the business content, arouse passer's imagination.
2. The central site over against axis open square of the hotel, not only continued centripetal force of Moon Bay's master plan, and became a main force to attract hotel guest.
3. Placing a sunken plaza at the end of the site, with quiet environment and excellent landscape, where placed with high—end restaurant, and formed three sequences rhythmically.

Correspondingly, in catering pattern, the district from the entrance to central square is lively due to close to entrance, where set western food, fast food restaurants are set to satisfy business men's needs. While the district from plaza to the ends which is far away from crowds and present with peaceful atmosphere, where Chinese food and specialty cuisine restaurants are set to attract

COMMERCIAL BUILDINGS | 商业地产

high—end business banquets. On the basis of type of business, the building volume presents with the distribution of large—medium—large, with its internal bay echoes to size; in space, because of its long and narrow shape, lacking of external resources, the designers gave up the idea of low—end store, and considered the gathering crowds from indoor space; In form, they adopted sheltered two layers' platform from the entrance to midpiece, and the bottom was overheaded, created a space that gathering popularity.

Architectural Design

The building used the technique that combined with stone, aluminum plate, glass curtain wall as standard elevation basic module, and work in concert with the whole building groups in Moon Bay and the across street hotel. The layout of this project formed staggered volume by advancing and retreating of front and back, which increased interestingness of the building; the establishment of billboard become one of the specialties of future business; remain margin for individuality exertion; The three floor volume in both side of inside street supplemented by glass curtain wall, which strengthened a sense of existence of east to west axis and south to north; While the sunken plaza with colored glass curtain wall and landscape stair highlights itself in the whole area. The design unite base shape, started from type of business, planed reasonably, focused on the important parts, controlled and coordinated with other parts, made well preparation for future individual exertion.

www.bacdesign.com.cn

住宅景觀・私家庭院
Residential Landscape・Private garden

酒店景觀・城市公共景觀
Hotel Landscape・Urban Planning Design

旅游度假項目規劃
Resorts and Leisure Planning

建築景觀模型
Architecture model

長期誠聘景觀設計人才,誠邀專業人士加盟.

地址: 廣州市天河區龍怡路1號農機物資公司綜合樓東梯六樓(郵編:510635)
Add: 6th Floor, Building of Agri-Materials Company, No.91, Longyi Rd., Tianhe Dist., Guangzhou(P. C. 510635)
電話Tel: 020-87569202 (0)13688860979/13360588555 Email:bacdesign@126.com